Ultimate A

Hot Rod

Timothy Remus

Published by:
Wolfgang Publications Inc.
PO Box 10
Scandia, MN 55073

First published in 1998 by Wolfgang Publications Inc., P.O. Box 10, Scandia, MN 55073, USA.

ISBN number: 0 9641358 8 4

Printed and bound in the USA

On the cover: The Deuce with the blown big-block is the work of David Tallant, a man who likes his '32 Fords red hot.

Ultimate American Hot Rod

Acknowledgements

After ten years and over thirty books, the idea becomes more and more obvious. If you distill it all down, my true role in this business is to act as the intermediary between professionals who weld, fabricate and paint, and the vast numbers of non-professionals anxious to learn how to weld, fabricate and paint.

Which is just another way of saying that I couldn't do all these books without all the "guest editors" who appear in each issue.

The list of guest editors who helped with this book is as long as my arm. Short interviews seem like an especially good way to bring to the reader the collective wisdom of men with years of experience building hot rods. Men like Pete Chapouris, Roy Brizio, Gary Harwood, Gary Schmidt, Kenny Gollahon, Greg Ducato, Gary Heidt and Ralph Lisena.

No less important is the help provided by Jim Petrykowski at Metal Fab, Dennis Overholser at Painless Wiring, Art at Morrison's, Ray at Currie, Gary at Dutchman and Brent at Fat Man Fabrication.

Many of the same companies supplied catalog images used to illustrate this fine edition. Thus I am additionally indebted to the staff and ad agencies of Bitchin Products, Chassis Engineering, ECI, TCI, Currie, Dutchman, SO-CAL, Heidt's, Harwood, Downs, Posies, Morrison's, and Fat Man.

The last two chapters document the construction of two very different cars. The very nice images of the Steve Moal roadster are the work of Michael Dobrin, a man with the eye of an artist and the heart of a true enthusiast. The photos of the Model A are provided by Eric Aurand - the car's owner and primary assembler. This means Eric often had to set up the shot with the camera on a tripod, click the self-timer and then jump into the photo. I also need to thank Steve Moal, another artist, for making the car available to Michael Dobrin and myself, and for help in explaining how and why he built the car he did.

I'm also indebted to a number of old friends for help in various forms. My old partner Steve Hendrickson from Rodder's Digest provided proof reading and advice, Bob Larsen (aka Yup) proof read the manuscript and Linda Jensen helped with the layout work. Of course I must close with a hearty thanks to my lovely and talented wife, Mary Lanz, for proof reading and moral support.

Introduction

What it is

Building a complete car of your very own is a large undertaking. Before you can start the construction you need to do some buying. But to be an intelligent buyer you need first to understand the various systems that make up the car. Only with a good understanding of how things work can you decide which parts and systems are best for your project.

How stuff works

As an aid to understanding I've included nine chapters that discuss everything from frames to chassis, to brakes, to alternators.

Each of these chapters attempts to explain how things work, why they work and some of the differences between similar parts or systems. Why, for example, you might want to run a straight axle on one car and an independent on another. Or what are the advantages and disadvantages of running one of the newer 700 R4 overdrive transmissions in place of the tried and true TH350.

Assembly sequences

For a look at the real world of busted knuckles and cross-threaded bolts, the final two chapters follow the assembly of two very different cars. The Steve Moal roadster, built for a certain famous TV personality, is hand-built in the extreme. Coach builders of old created a lovely hand-crafted body designed to fit on one of the factory chassis of the day. Steve takes the concept one step farther, by creating a beautiful hand-crafted body designed to fit an equally interesting hand-crafted chassis.

And lest you accuse me of presenting only high-dollar cars built by professionals, the final chapter documents the building of a very simple, early-style Model A. This second hands-on chapter starts at the very beginning, with a picture of the old Model A body hanging from the rafters of a barn in South Dakota. The final sequences were photographed in a small garage in Iowa where most of the assembly was done by Eric Aurand with help from his father.

The two project cars represent very different ends of this still-evolving hobby and industry. They are presented here to show how each was constructed. And to show that hot rods and street rods don't have to fit some boiler-plate format developed by the guys at the local car show, or the magazines or who ever. Whether your own dream car is a Deuce with fenders, or a dare-to-be-different Nash, doesn't matter. Use your imagination, your skills, and some careful planning to design and build the car that *you* really want.

Chapter One

Planning

The Most Important Part of the Project

WHO NEEDS A PLAN

Whether your hot rod project is simple or complex, you're going to need to work out a plan. Before turning a wrench or spending the big bucks on components and tools, you need to know exactly what you're building and why.

No matter how careful you are or how many parts you buy at the swap meet, building a hot rod is never a cheap proposition. Even a beginning street rod kit based on a 'glass body and simple frame, coupled with simple drive train options from the local junk yard will often run past ten thousand dollars in short order.

Eric Aurand, owner of the Model A seen in Chapter 11, is an artist by training. Before building the car Eric did a series of sketches and then the rendering seen here. "I was going to build it without fenders" says Eric. "But after looking at the sketches I decided it would look much better with the fenders."

The point isn't just money and how much you intend to spend. A good plan for the car and more specifically the body of the car will do at least two things. First, it will help you spend your budget wisely. By planning the project and doing a budget you will be forced to make logical decisions for all the components and avoid the spur-of-the-moment purchases of the latest new and trendy trinkets. Second, if you really want to make a statement with this new car, if you really want to build a car that has impact, then you need a good plan so everything on the car fits together and adds to the car's visual impact.

Great street rods that end up as "cover cars" don't happen by accident. Take the time to do a plan and then stick to it during the construction. Kelsy Petrykowski

Some cars at the rod run really stand out. Not just the expensive ones but rather the cars where everything comes together with a certain harmony. It might be a simple Model A like the one seen farther along in this book, or a professionally built Deuce. The point is that everything fits in a visual sense and contributes to the overall good looks of the car.

Great cars don't just happen, they are planned. We all like to think of ourselves as craftsmen, able to assemble a car with perfect door gaps and the flattest of panels. Just as important as the precision and craft, however, is the design. You want a car that has an impact on the viewer. A design that looks as current in five or ten years as it does the day it rolls out of the paint booth.

The plan should include a budget, a sketch of the car, and the dimensions of the body and chassis. As mentioned earlier, the budget will force you to do some good accounting as to what you really can and can't afford.

Note the nostalgia theme: Steel wheels, white wall tires, drum brakes, a detailed flathead, minimal chrome and an early "flamed" paint job that extends right down to the frame. Kurt Senescall

Nostalgia Machine or Cutting Edge?

You need to decide first what style of car you're going to build. Will it be a near stock resto-rod,

You don't have to chop the top but you do need to think about proportion, design, and good taste. Kurt Senescall

There are no rules here, only that the car you build look like a hot rod when you're done - and this smoothy certainly does. Kurt Senescall

This early-style hot rod relies on five spoke mags, primered paint and a Chrysler Hemi to lock in that old-time religion. Eric Aurand

or a very contemporary smoothie? Remember that you are in fact designing a complete car - so it's in your best interests to take your time and give this some thought.

While you don't have to do what's been done before, you need to try and keep things in proportion and follow some semblance of good taste. Find a style that you like and try to keep the entire car consistent with that style. Directional billet wheels, for example, would look odd on a resto-rod. Remember that every part on the car contributes to the whole.

In order to better clarify your ideas on this design business it's a good idea to start two photo files. If you're building a Chevy coupe, collect photos of Chevy coupes, both those you like and those you don't. Spend some time studying the photos (taken at shows or clipped from magazines) until you understand what it is that gives certain cars their great impact.

Professional shops rely in the skills of designers like Chip Foose or Thom Taylor. The designer usually works up a series of sketches until the customer says, "yes, that's the car I want." The sketches are then formalized into a rendering of the car, usually done with color and all the correct proportions. This rendering works to show the customer what it is he or she is buying, but it also works to keep all the men in the shop focused on the same car. Everyone knows what that car is supposed to look like, there is no guess work.

For the non-professional builder working at home there are a couple of options to this rendering business. First, you can hire someone like Thom Taylor or Eric Aurand. Compared to the cost of the whole project, the cost of hiring an artist becomes insignificant. Your second option here is to start with a photo of the car, make a series of enlargements, and run off a series of inexpensive copies. Now you can experiment with a scissors and glue. In this way it's easy to see the effects of a chopped top or channeled body. You can even buy colored markers to try out some different colors.

Computer-capable builders can do the same thing by scanning a photo of the car into the computer and then doing the cutting, pasting, and painting with the software of their choice. Whether the approach is low-tech or high, the important point is to start with a nice side view that doesn't distort any of the car's dimensions or proportions.

You need to know what the car is going to look like and what the body dimensions will be before you start cutting and welding. If you do your planning from a photo you can take dimensions off the photo. After pasting the photo back together with the top chopped just right, you can measure relative distances off the photo and transfer those proportions to the new car. For example, if in the photo the distance across the door (something you can measure on the car itself) is five times the dimension from the top of the door to the top of the roof then that relationship will hold for the actual car.

No matter how sophisticated your sketches are they will help you focus on the car's design, something that more builders should do. The finished sketch or rendering will keep you focused once you start building and help you avoid new trends and distractions that happen before your car is finished.

The sketch should include dimensions of the car, including chassis dimensions. You need to know the wheelbase and the width of the track, front and rear. The distance from the frame rails to the ground, "ride height," is another important dimension that should be determined before the building begins. A separate sketch of the chassis is a good idea. One that shows the engine in relation to the radiator and firewall.

If you are starting with an existing car, park it on a level surface and measure everything before it changes. Overall length, bumper to ground heights, frame to ground clearance. All these measurements will give you a set of basic measurements, a starting point for the project.

And though you need to know the exact dimensions before and after any changes are made, most designers stress the importance of proportion, not dimension. To quote Thom Taylor: "Sometimes when they chop a top they go really wild and cut the top five or six inches. So now they've really changed the proportion between the top and the body and some of them look like they flipped over and landed on their top. If the builder is after a Bonneville or radical look, that's fine - if that's really what they want. There isn't anything bad, but the cars that everybody likes tend to have good proportion and a certain continuity in all the components."

Ultimately, a plan is just a means to an end. You want a particular car with a certain set of features. The plan is just your insurance policy. So that after the all the fabrication and sanding, and the application of the perfect paint job, the resulting car really is the car you set out to build in the first place.

In this chapter I've tried to emphasize the need for planning, as a way to create a car that's more than the sum of its parts. During an interview with Greg Ducato from Phoenix Transmission Products (found in Chapter Seven), he kind of hit the nail right on the head as we talked about the mistakes people make when they build a car. Greg's comments from that interview bear repeating here, as a way to sum up all the reasons for making a plan.

"There's so much stuff out there that it's easy to get distracted by the options and lose the focus you had for the car. People read about this guy or that guy and his car in a magazine, but the magazines don't do a good job of really describing the technical side of the car and they also don't tell you if the builder has had any trouble with the car or had to change the rearend ratio to make the engine/transmission package work correctly......"

I tell people, "Don't get carried away with power, these are light cars to start with... Stick with reliable set ups instead of the exotic stuff. Keep sight of the 'big picture.' Tire size and gear ratio are just as import as the color you paint the car, but a lot of people miss that... And you've got to build the car for yourself, don't built it to please the judges or your buddies."

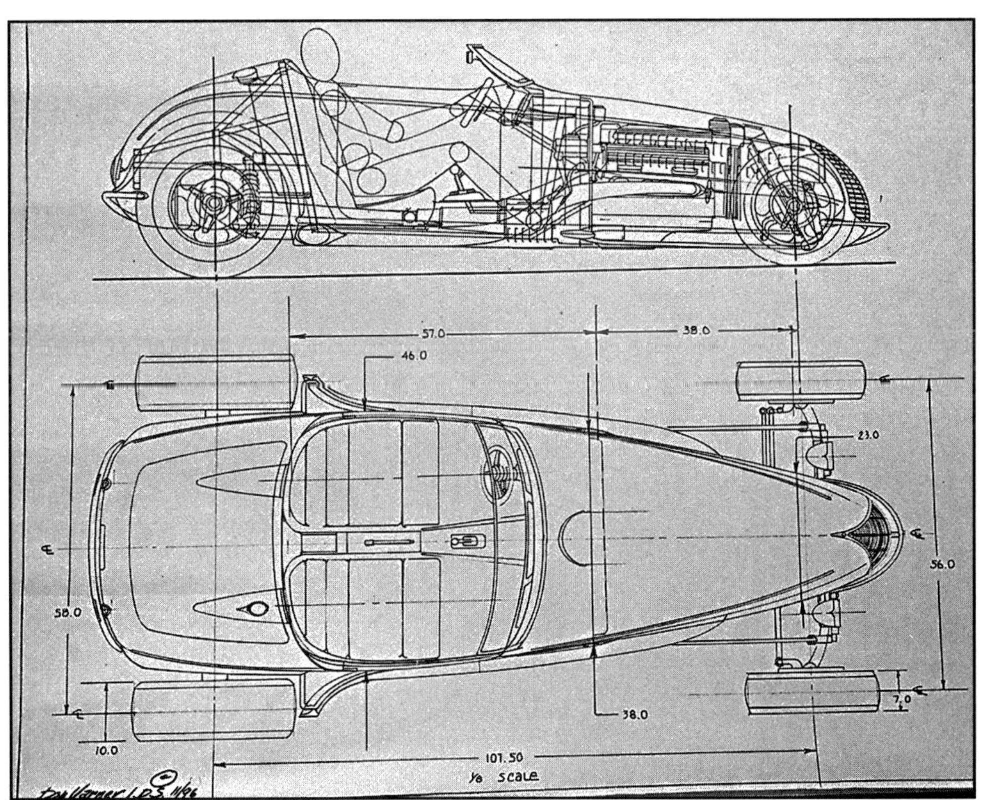

Steve Moal, builder of the roadster seen in chapter 10, likes to work from a very complete set of blue prints that show everything, including the position of the engine, drivetrain and the driver.

Chapter Two

The Right Frame

One Decision That Affects All Others

Though it might be a cliche', the frame is in fact the foundation upon which your car is built. The frame will also affect and be affected by your decisions regarding the overall design for the car. Obviously a pro/street car will require a different frame than one that runs more modest rubber. Even without the huge rear tires a car with a massive motor requires a frame that is extra stout and better able to handle the additional torque of a big mill.

When it comes to picking a frame there are at least

Complete frames come in every size and shape, including this "step boxed" Deuce frame with Pete and Jake's ladder bars in back and a dropped axle with radius rods in front. Frame rails come from American Stamping, cross-steer set up uses a Mullins/Vega steering gear. SO-CAL

three options: Use the frame that came with the car; buy a frame from an aftermarket company (with or without suspension attached); or buy a set of rails and build your frame from there.

But first a look at the various frame designs

Even Mopar fans can find complete frames like this, designed to fit most '33 and '34 Dodge and Plymouth models. Frame comes boxed front and rear, with X-member, leaf-spring rear suspension, and independent front suspension already installed. Chassis Eng.

Basic designs

Most of the early frames were based on a simple rectangle, with a few simple crossmembers to tie the side rails together and support the various suspension components. A rectangle, however, doesn't have much structural integrity. Early designers and engineers discovered that a few triangles placed in the center of a frame (commonly called an X-member) did a tremendous job of reinforcing the basic rectangle without adding too much weight.

As time went on more and more of the early automobiles got X-members in the center of their rectangular frames. This was done to help those cars deal with the additional weight and horsepower that came along with the other advancements of the automobile. The side rails became stronger too, becoming taller, going from the three or four inch tall rails of a Model A to the five or six inch tall rails of the '46 Ford. As the side rails became taller they tended to be reinforced and partly "boxed" in order to add even more strength.

Today a street rodder building a frame can follow the same evolutionary path used by early car builders and designers. As the weight and horsepower of your car increases the frame may need to be reinforced, often by the same means used many years ago.

Unless your car is a very early model, like a Model T, it's likely the original frame can be

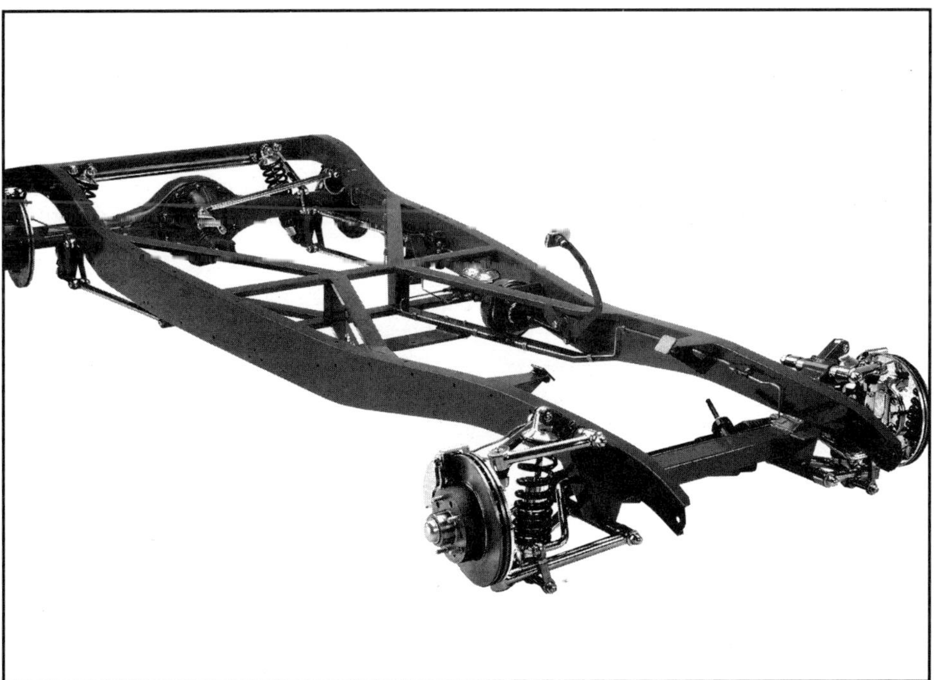

Frames can generally be ordered in raw form or quite complete. This '33, '34 Ford frame is equipped with the deluxe front suspension kit, rear disc brakes and stainless brake plumbing. TCI

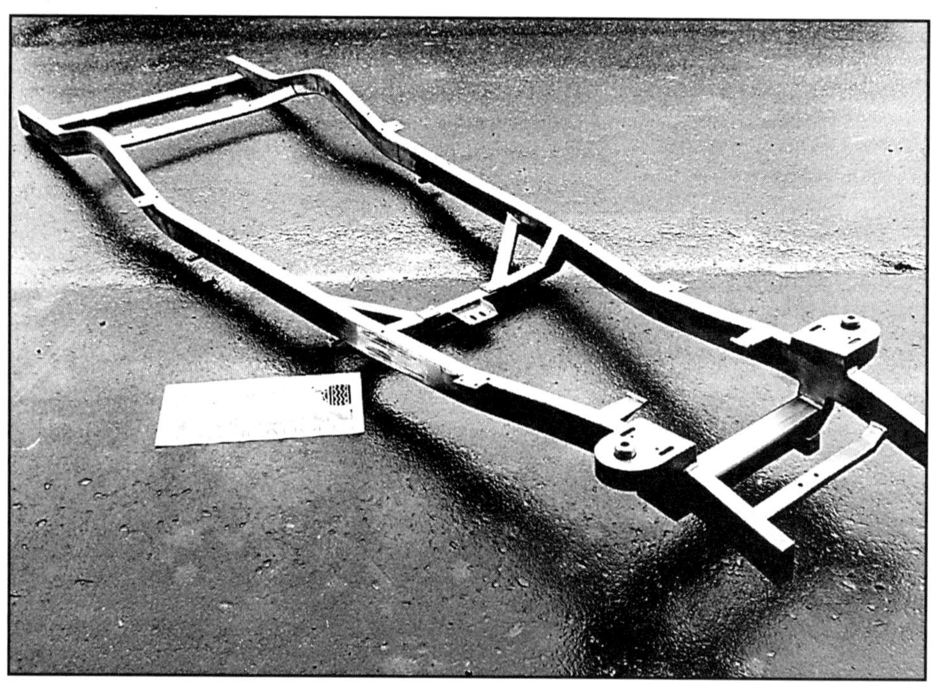

Bow-Tie fans can order this frame for '37 - '39 Chevrolets. Rails are formed from .188 inch mild steel, 2X4 tubing mandrel bent for smooth, accurate bends. Comes with a Mustang II cross-member in front and mounts for either big or small-block engines. Available with front and rear suspension installed. Fat Man

used as the basis for your new street rod. Before deciding to stick with the original frame, however, take the time to do a thorough inspection. Look for rust, serious rust, deep and pervasive enough to weaken the frame. Look too for cracks in the rails and cross-members, and damage from any old accidents or abuse. If there's doubt as to whether or not the frame is straight, set up the bare chassis on a surface table or frame jig and check each corner for height and measure diagonally across the frame to ensure it's still square. Sometimes you're money ahead to simply scrap the old one and buy a reproduction frame.

Model A and Deuce frames will likely need an X-member added to the center of the frame. Most GM frames from the 1930s and 1940s came without X-members, so you may want to reinforce these frames with a central X-member or additional cross-member depending on your intended use.

Even the better Ford and Mopar frames with X-members may need modifications and reinforcing to handle modern loads and suspension components. Un-boxed rails, like those used on Deuces, can be boxed over all or part of their length. Pre-cut boxing plates are even available for many of these frames. Boxing should be done with material that is the same thickness as the frame itself, probably .120inch plate. Frames need to be boxed in the center where any X-member will mount and all the way forward, past the point where the engine mounts attach to the frame. When in doubt, or in the case of a high-horsepower car, you probably want to box the entire length of the rails.

The X-member used by

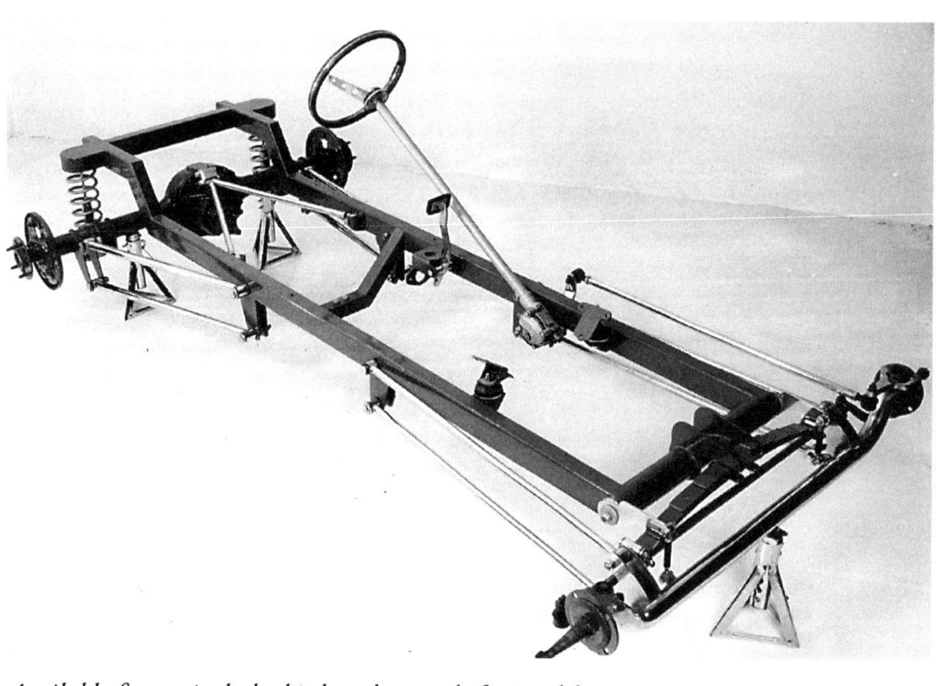

Available frames include this boxed example for Model Ts. Frame comes complete with mounts for the engine and transmission of your choice. Designed to accept VW van steering column and box. Comes with or without rearend assembly. Chassis Engineering

Henry and others adds a great deal of strength, though it often gets in the way of the tail-shaft housing when automatic transmissions are installed. A TH350 will usually require only some light notching of the X-member, while a 700 R4 may force you to do more extensive surgery on the X-member before it can be installed in an early Ford frame. Kits available from a variety or sources make it easy to create more room for tail-shaft housings without losing any of the strength built into those Ford X-members.

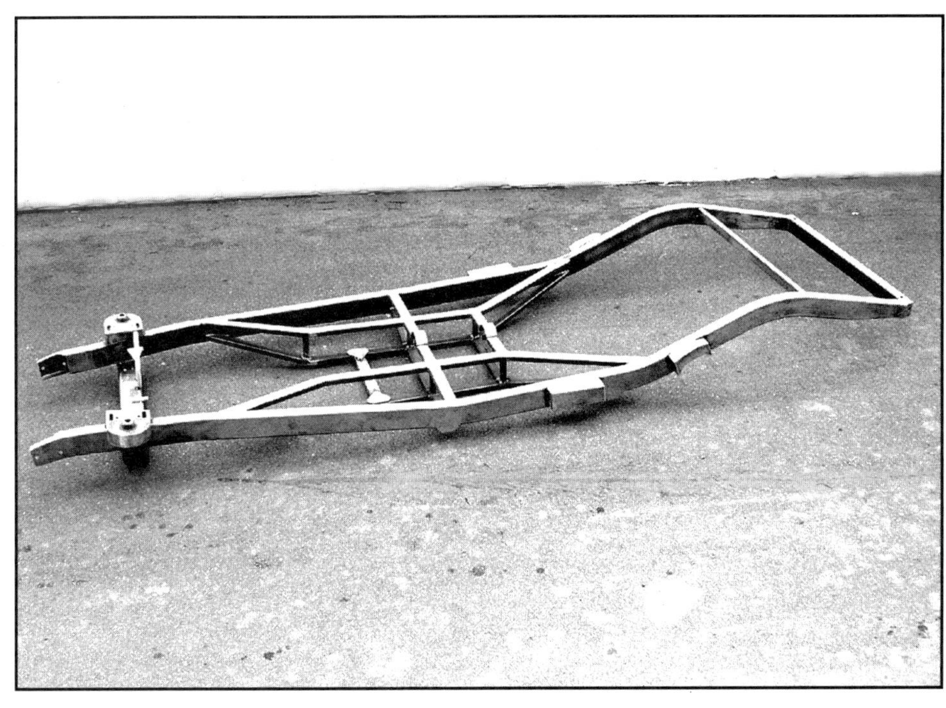

This '35-'40 Ford frame comes with Mustang II cross-member and a tubular X-member already in place. Fat Man

Buy a complete frame

If you don't have a frame, or the existing frame is too old and degenerate, then it may be time to buy a complete frame from the aftermarket. Eric Aurand, who built the Model A seen farther along in the book, says he left the original frame lying in the weeds when he went out to South Dakota and bought the A-bone. "For what it costs for a new reproduction frame," explains Eric, "it doesn't make much sense to clean, repair, and reinforce the old one."

Options here range from a bare frame purchased from a small shop like Metal Fab outside Minneapolis, Minnesota, to a complete frame with trick independent suspension from a company like TCI, or Chassis Engineering. When it comes to frames you can have anything you want. Pro/street, with or without suspension, or with the new air-suspension system from Art Morrison and others

The complete frame comes with a number of advantages. First of course is the obvious fact that new is new, which means you somehow have to get along without rust. New frames often have other more subtle advantages as well. Like

Some cars have a frame which doesn't allow a simple Mustang II cross-member installation. For these situations a complete frame clip might be the answer, available for various Buick, Caddy, Chevy, Dodge, Mercury, Ford, Nash, Olds, Packard, Plymouth, Pontiac and Studebakers. Fat Man

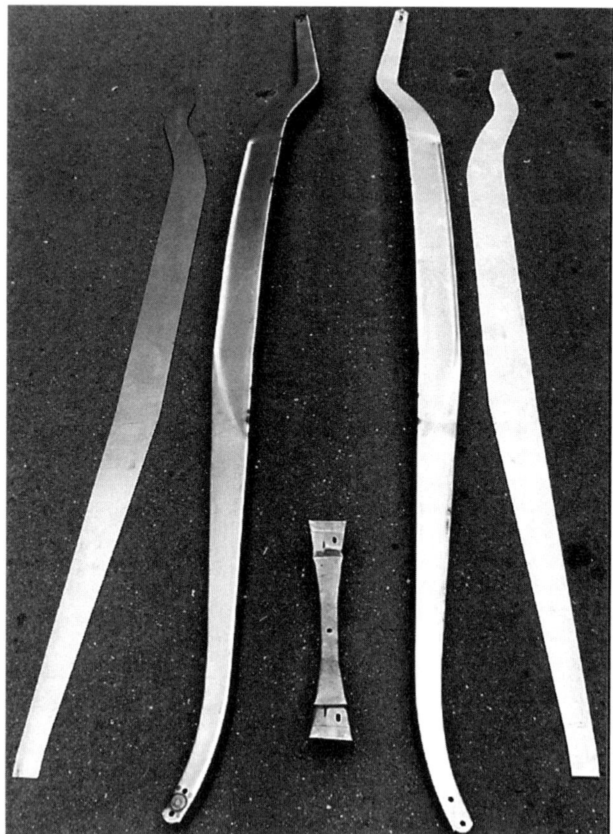

The start of a Deuce frame, stamped rails from ASC that measure a full six inches in the center and are heavier than Henry's. Boxing plates are from ASC as well.

better reinforcing with boxing plates already installed and stout X-members where the original frame might only have had one or two simple cross-members. Whether the original frame had an X-member or not, the X-member on the new frame probably allows more clearance for installation of an automatic transmission.

BUILD A FRAME FROM A SET OF RAILS.

Despite the fact that complete frames are available, street rodders often build their own frames from a set of rails. The advantages of "doing it yourself" include cost. By purchasing a set of rails and fabricating your own cross-members, or buying the cross-members too, you're sure to save money. Some professional builders start with rails, that way they don't have to worry that a complete frame was bent ever so slightly during shipping. Finally, by building your own frame you can have whatever you want, be it a pro/street frame or a longer wheelbase, or rails that are kicked up in the rear.

Rails and cross-members

Most aftermarket rails are made from mild steel, flame-cut or laser-cut and then welded together in a jig. Some rails are stamped, like the Deuce rails from American Stamping Company. The accuracy of these rails is dependent on the jigs and dies, and the skill of the men and women who do the cutting and welding. Not all of these components are the same quality. Some are built with open channel while other use boxed tubing or even round tubing. Look and compare the various offerings on display at the shows. Then ask professional builders which set of rails *they* prefer and why.

The frame construction takes place at Metal Fab outside Minneapolis, MN. The surface table was leveled out first with adjustments on each leg. Then the frame rails were set in place at ride height.

Cross-members and complete X-members are also available to complete the construction of that new frame. The range here is substantial. The catalogs list everything from replacement X-members for Ford frames to new X-members intended to reinforce frames that never had one to start with.

Many frame rail manufacturers will do special orders, like a longer than stock wheelbase or a "pinched" Deuce frame for a Model A body. When you call and order the rails and other frame components buy the best quality you can find. The more uniform and accurate the rails and cross-members are the easier it will be to construct a complete frame that is straight and square.

Frames and cross-members are generally made from mild steel, at least .095inch thick. As the car gets heavier and the loads increase .120inch mild steel is preferable. Chrome moly tubing is generally only used in round-tube competition frames.

The frame jig or surface table

Before you can build a frame you need to construct some kind of a jig to hold everything in position. The sketch you did early in the project will ensure that you understand what you want. The frame jig or table will help ensure you achieve those goals.

Boxing plates are welded to the rails in a series of ever expanding tach welds so as not to put too much heat in one area. Threaded inserts for body mounts must be located inside the rails before the boxing plates go on.

Most professional shops that construct frames use a surface table, a table with a perfectly flat and level surface. A table like this will allow you to set up two rails that are both at the same *exact* height.

A portable surface table can be built with bolt-on legs. The important thing is that the top surface is as level as possible, that the table be very sturdy and that each leg is equipped with a height adjustment so the whole table can be leveled out. That way you have a known reference point during the frame construction.

You need a sturdy stand at each corner of the new frame. How elaborate these stands are is up to you. Some builders just use square tubing and tack weld everything in place once the rails are in position. Just be sure the frame rails can't move once they are bolted or tack welded to the stands.

Spreader bars at the front and the rear are from the Deuce factory. Pete and Jakes front cross-member is designed to drop the front of the car. All welding is done with a heli-arc.

Once you've got a table

Nuts & Bolts

More than meets the eye

Ultimately, what holds your entire hot rod together is a bolt, or a bolt and nut combination. Bolts are not all created equal, so it pays to understand the differences between the good bolts and the cheap, you-don't-want-them-holding-your-car-together, bolts.

All bolts are graded. The best are grade 8, followed by grade 5 and then grade 3, sometimes called hardware-store grade. The best bolts are made from forged material and feature rolled threads. Cheap bolts are made from non-forged steel and have threads that are formed by a cutting or smashing operation. Allen bolts are nearly always a grade 8, it's simply a standard set by the manufacturers.

In any given size, a fine threaded bolt is stronger than a coarse thread bolt for two reasons: The fine thread bolt has a larger root diameter (the diameter at the base of the threads) and there is more physical contact between the threads.

Some builders love Allen head bolts, They should, however, understand that there are a few drawbacks. Like the small head size which means the bolt can't clamp two parts together as effectively as a hex-head bolt unless you use a washer under the head. These bolts are so tough a standard washer will deform under the head of a very tight Allen bolt, and after it does the bolt will no longer be tight. Use hardened and ground washers under the head of those Allen bolts, especially the ones that hold important things together.

Shiny bolts are obviously better than plain old steel bolts. Chrome plating a bolt is a good-news bad-news deal. The bad news first: chrome plating weakens a bolt slightly. The good news: the bolts they plate are nearly always at least a grade 5 and often grade 8.

The other bad-news part of the chrome bolts story is the buildup of material on the threads which can make the bolts slightly oversize. This possible oversize means a chrome bolt can gall in the hole. Now, if a cheap bolt gets stuck in a hole there's some chance you can get it out. With a grade 8 bolt, however, you can hardly begin the process because you can't even get a drill bit to penetrate the busted bolt.

This lengthy preamble means that among the "must-have" tools stashed in the garage is a set of taps and dies. Use these to chase the threads on all the chrome bolts and all the holes they will thread into. In fact, use these to chase any hole or bolts you have doubts about.

Anti-seize or Loctite on the threads will also help prevent a busted bolt because either product will prevent metal-to-metal contact between the male and female threads.

Though they look pretty, bolts made from stainless steel aren't as strong as you might think. Stainless bolts tend to stretch and distort more than steel as they are tightened, for this reason they should only be used once and some kind of thread lubricant or anti-seize is a good idea. Even the stainless bolts rated as grade 8 are suspect due to the fact that the grading system is different: A grade 8

Six dashes radiating from the center indicates a grade 8 bolt, while three dashes designates a grade 5. No dashes mean it's a low strength bolt of unknown properties. Allen bolts are nearly always a grade 8. The bolts they chrome plate are often grade 5 or 8, but it's hard to tell after the head is polished.

stainless bolt is only equal to about a grade 3 steel bolt.

When looking for bolts to assemble that new hot rod, consider that some bolts are more important than others. The 1/4 inch bolts that mount the gas tank to the frame aren't under the same load as the bolts that locate the brake caliper to the caliper mount. In the case of the brake caliper, always try to use the bolts that come with the installation kit or a grade eight substitute. For structural, important bolts, use the best you can buy. If they don't come with the kit then source them from your local hot rod shop, a good industrial supply house, or even the local farm implement dealer.

Take the time to read the marking on the head of the bolts and read the label on the box. Look for *Grade 8* and *Made In USA*. Avoid anything labeled "better than grade 8," they're usually counterfeit bolts of mediocre quality.

A variety of means are available to prevent these expensive bolts from backing out of the hole. Split ring lock washers, Nylock nuts and at least two grades of Loctite are among the most popular of the options here.

Loctite: red or blue?

Loctite makes a variety of products often used to prevent the loosening of nearly any nut or bolt. I grew up in a two-Loctite world. There was blue and red (never mind the numbers or labels on the little bottles). Blue was used for most of the common situations encountered in an automotive shop, red was the stuff you used to absolutely guarantee that not even an act of God was going to shake apart the bolt.

It turns out there are at least seven different threadlockers from Loctite. What we call blue Loctite is technically a number 242 or 243. Either one is considered by the company to be, "a medium strength threadlocker for fasteners up to 3/4 inch." Number 243 has a slight advantage in that it's slightly stronger than 242, is quicker to set and is more tolerant of a little oil on the threads. Both products protect against rust and corrosion and both allow removal of the bolt or nut with hand tools and no heat.

The Red Loctites include numbers 262, 271, 272 and 277. These are indeed stronger than the blue products, but before reaching for the red material consider that some of these are meant for parts that will not be disassembled, (number 270) while the rest of the redtites "require heat and hand tools for disassembly."

In order to work properly and form a good bond Loctite requires an oxygen-free atmosphere and active metal ions. Aluminum parts are a problem as they offer the Loctite no free-metal ions to aid the locking process. Chrome plated and stainless bolts have the same problem. All of which means that when using plated bolts or when screwing a bolt into aluminum it should first be treated with Loctite primer, either the T (which makes for a stronger bond) or the N (which can be used in colder weather).

And despite the fact that the 243 will tolerate oily threads better than some other products, it's always a good idea to clean any dirty threads with Loctite's own Klean-n-Prime, or a product like Brake-Klean, something that leaves no residue behind, before applying the Loctite.

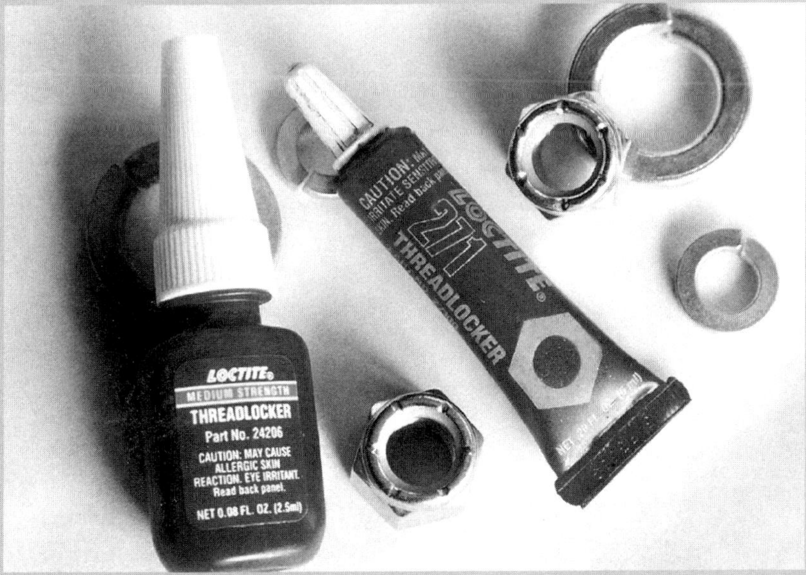

What we call red or blue Loctite is actually a wide variety of products, from medium strength thread lockers to high strength loctites that may require heat to loosen.

Solid struts are used in place of the shocks to keep the frame at ride height. The frame will be "C'ed" above the axle so it can ride lower.

and some stands, you need to set the two rails up at *ride height.* The distance from the bottom of the rails to the table should be the same as the height of the frame rails to the ground when the car is completely finished. All the work done on the frame, including the installation of the suspension (minus the springs), should be done at ride height.

You can't just lower the car after it's built as all your planning, not to mention suspension geometry, will go down the tubes. It's much better to design and build the car at the correct ride height. Your initial drawings and the information provided by the rail manufacturer is essential in setting up the frame rails, getting everything welded together and attaching all the suspension components.

No matter who made your frame rails you will need to determine a few necessary reference points like the frame centerline and the centerline of the front and rear axles. Most manufacturers of reproduction chassis and chassis parts provide diagrams that show the location of body mounting holes, cross-members and centerlines. Original factory frames came with their own reference points shown in many manuals and books.

Frame Building and Checking

Some professional builders like to use spreader bars designed for the car, that they know to be the correct dimension, as a good way to jump start the frame-fabrication project. With two rails and the spreader bars in place at either end you have the basic frame intact.

Once the rails are positioned roughly at ride height with spreader bars clamped or bolted in place, you should spend some time with a tape measure and a level, making sure that everything is perfect. More specifically, make sure that the two corners at either end of the frame are at the same exact height, and that the level set on a cross member or across the two rails is indeed level.

To make sure the frame is square get a helper and measure corner to corner in crisscross fashion. These measurements must be exactly the same. Use two different mea-

Almost finished, Jim will put a dummy engine in place before locating the engine mounts. Three-bar rear suspension is finished. Clutch linkage will be hydraulic.

suring points as a good way to double check your measurements. It's also a good idea to check that each section of the frame is square as well. If the two diagonal measurements come out the same then you know the frame is square.

Jim Petrykowski from Metal Fab warns that, "A lot of builders use the body mounting holes as reference points and that's not a good idea. On some cars those holes aren't very accurate. People should use reference holes in the frame, like the "suspension bumper" holes in Ford frames located at the axle centerlines. These holes were used to accurately locate the frames as they moved down the line at the Ford plant.

If builders use body mounting holes in the top of the frame they must be sure that all the holes are the same distance from the edge of the rail. You might simply want to make your own reference points, along the axle centerline at each corner. After all, you're going to have to determine axle centerlines anyway. Just be sure each reference point is the same distance from the centerline of the frame.

When you're done you should have a basic frame sitting on the table at ride height. The new/old frame should be level from side to side and square from corner to corner.

Building a quality frame relies mostly on careful set-up and preparation for a good result. With all the rails set up correctly on the table the rest of the cross-members can be welded or bolted into place. Diagrams from the factory or the manufacturer will be your guide as to the exact placement of cross-members. As you weld in the cross-members be careful that the rails don't move. A T-square set on the table can be run along the rails occasionally to make sure the rails haven't begun to twist.

The side-to-side measurements through the center of the frame need to be checked as the frame fabrication continues. A porta-power can be used to push the rails out slightly and a carpenter's clamp used to pull them together. Remember that welding creates heat and heat makes metal move. Check the dimensions often and try to move around as you weld so all the heat isn't concentrated in one area. As always, patience and persistence are the keys to quality.

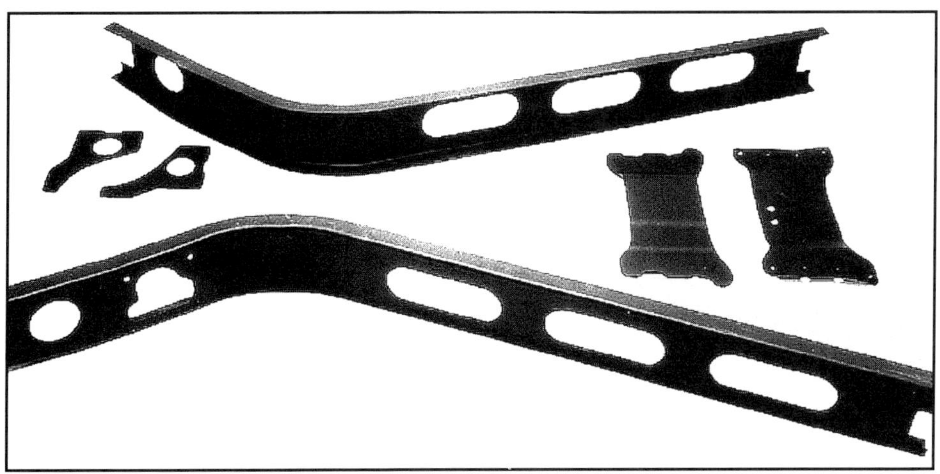

X-members can be added to many frames with kits like this. Most are designed to provide extra clearance for the transmission tailshaft housing. Chassis Engineering.

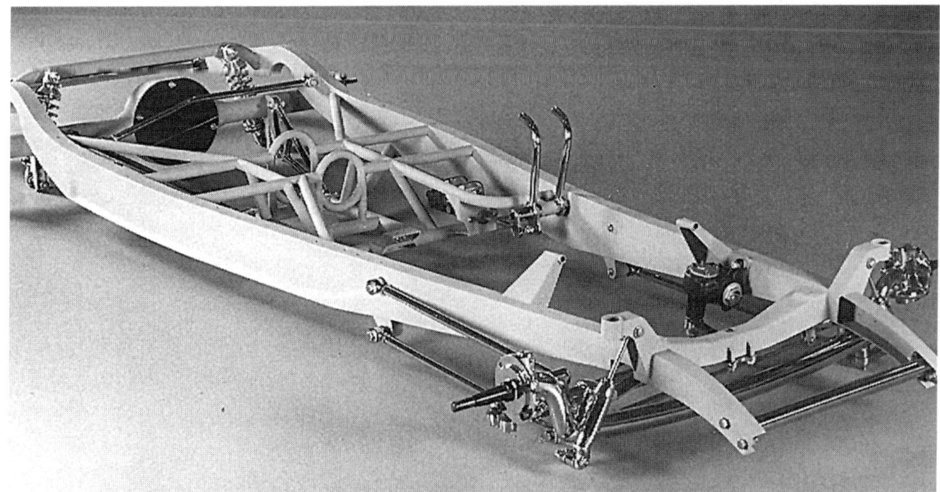

This Deuce chassis from Harwood uses boxed rails, raised motor mounts for good ground clearance and the ability to run a mechanical fan and tubular X-member so exhaust does not hang below the frame. Comes with or without: front and rear suspension, axles, clutch linkage. Harwood

Chapter Three

Front Suspension

Form *and* Function

How independent do you wanna be?

The types of front suspension systems available for your street rod range from simple straight axles to more complex independent suspension systems. Some come as kits and use a variety of OEM parts while others come as a complete sub-assemblies that include all the suspension components, complete brake assemblies and even the cross-member. The hard part is deciding which type of suspension to use, and then deciding how best to implement that decision.

Making the decision between a straight axle and independent suspension would be easier if the only con-

This front suspension features all polished components, including the anti-roll bar. Comes with adjustable coil-over shocks, power or manual rack and pinion, vented rotors and four-piston brake calipers. TCI

This is the first photo in a sequence showing the installation of a Heidt's front end in a stock '34 Ford frame at Jon Kosmoski's shop in Minneapolis, MN. Note the stock suspension, complete with wishbone and lever-action shock absorbers.

sideration were function or value. Hot rods, however, are soulful machines and the aesthetic considerations can't be ignored.

The straight axle vs. independent decision is affected by the car's style, age, and intended usage. Some early rods for example, just don't look right with independent suspension, especially if the car in question is a nostalgia machine. In reverse, a real high-tech car might not look good with a dropped straight axle. Ultimately you need to consider the style of the car you are building and then choose a suspension system that fits your budget and the design of the car.

The intended height of the finished car is part of this decision as well. Some designs get in the weeds more easily than others, and some allow nearly instant height adjustment.

The final consideration should be the quality of the parts and how well they fit. Not every aftermarket suspension system will fit every street rod. In particular, only a few of the independent systems will fit correctly under the front fenders of early Ford and Chevy street rods.

Straight axles

Straight axles come in various flavors, including split wishbones and four-bar set ups. Henry designed a simple and durable straight axle supported by a buggy spring and held in place by the "wishbone" that attaches to the axle at either end and then comes back to a common locating point between the frame rails. The wishbone kept the axle from moving front to rear and also provided lateral support.

Though a fine engineering concept, the wishbone got in the way when the early four and eight-cylinder engines were replaced with larger and more modern V-8s. Some early hot rodder cut off the wishbone at the common locating point and suddenly had two "wishbones." With a little heat from the magic wrench he soon had each wishbone located to the frame rail instead of the common locating point.

The split wishbone left

After pulling out the old flathead, the next step is to disconnect the stock Ford shock absorbers.

Here Jon disconnects the shock on the left side, next comes the brake line, the spring bolts and the wishbone.

more room for big motors but didn't do much for the handling or geometry. When one wheel of a split system encounters a bump or driveway it pivots on the long wishbone - experiencing a caster change as it does so - while the other wheel rides along without any change. What this does is place a twist on the axle as one wheel goes up and over a bump. For this reason a tube axle should never be used with split wishbone or hairpin radius rods.

To use a tube axle a system is needed that allows each wheel to move up or down without twisting the axle.

The four-bar setup:

The four-bar or four-link system uses two bars on each side to hold the axle in place. When one wheel encounters a bump it moves up, the bars act as a parallelogram-type linkage and there is no caster change and no twist on the axle.

The four-link system has a number of advantages when compared to its evolutionary predecessor. Because each wheel operates somewhat more independently of the other, the ride is improved and a given bump has less tendency to upset the entire suspension. The other advantage is in the appearance of the four-bar system; some people think a chrome plated four-bar setup adds to the visual appeal of the front end, though anyone building a nostalgia rod might feel otherwise.

Though an uncommon arrangement, a four bar system can be used with a coil-over at either end in place of the buggy spring. The coil-overs help to further improve the ride of the straight axle and make it easy to adjust the height or change the spring rates.

Without a wishbone the only thing locating the front axle side-to-side is the buggy spring itself. Thus most manufacturers insist on the use of a panhard rod to locate the axle side to side. Some like to see a sway bar used with a buggy spring.

Early non-Fords often used two long parallel leaf springs to support and locate the straight axle. The major disadvantage here is style; the system just doesn't look as good or traditional as a buggy spring. In a functional sense the parallel spring setup works at least as well as the classic buggy spring.

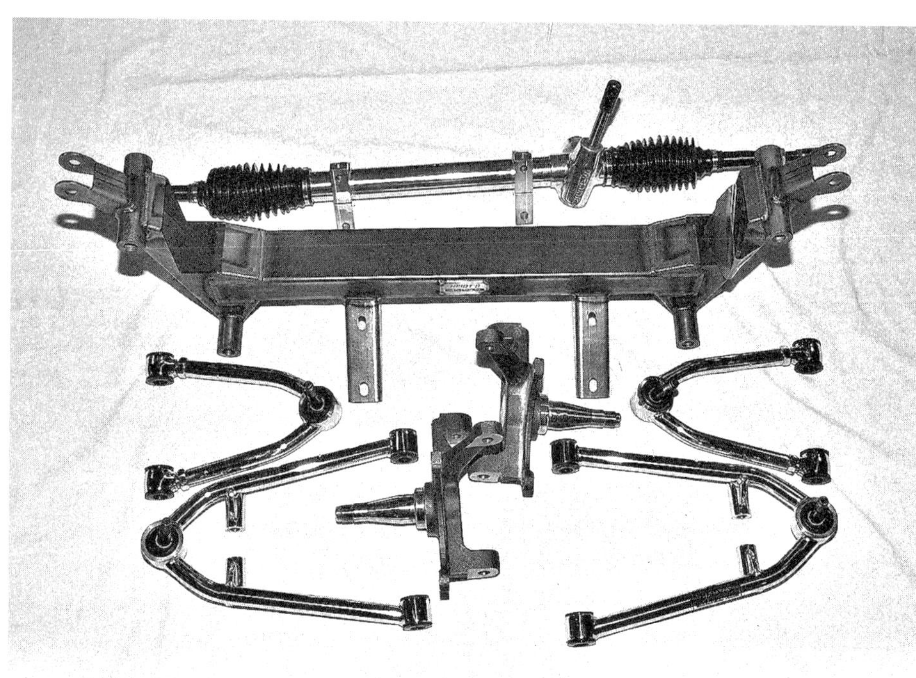

The kit being installed is the Heidt's Superide with polished upper and lower arms, spindles, and a rack and pinion gear ready to install.

Independent suspension

An independent suspension system offers dramatically improved ride by separating the movement of one wheel from the movement of the other. Each individual wheel can follow the road, *independent* of what's happening on the other side of the car. Each wheel is also able to maintain much better contact with the asphalt and achieve better grip and overall better handling. A good independent front suspension will also provide camber change in corners, anti-dive on braking and a low roll-center to improve handling.

Many street rod builders are installing one of the new air-suspension systems available for both the front or back of many cars and chassis. Companies like Art Morrison offer complete stand-along systems that include air bags, suspension arms and all necessary linkage. Other companies offer air bags and "shocks" designed to replace the springs in your current system. (For more information on air-suspension see the side-bar in Chapter Four.)

Compared to a straight axle, independent systems are more complex and expensive. As mentioned some hot rodders would say they just don't look right on certain early cars.

Most independent systems and some straight axles use an anti-roll bar. One wheel can't more up more than the other without twisting the anti-roll bar, thus the bar helps to minimize body roll. Anti-roll bar strength is determined by the size and make-up of the bar itself and the length of the levers on either end. Generally, as the bar gets bigger, it gets stiffer. As the levers get longer the stiffness goes down. Though often associated with independent front suspension, anti-roll bars can be used on various types of rear suspension, as well as straight axle front suspension.

Buying considerations

If you've settled on one type of suspension the next step is to find the manufacturer that offers the best example of the system you want. Beware suspension kits that claim to fit too many models and years. Try to find a hot rodder with a car like yours who can provide some kind of endorsement.

When buying a Mustang II style suspension, try to buy one that uses stock, Ford, geometry and a stock rack and pinion position. When the position of control arm pivots and the dimensions of the arms themselves are modified (to make them fit under the fenders of early hot rods for example) there can be surprising and unpleasant surprises.

Note: this chapter includes a side-bar on wheel alignment.

These dropped spindles are Heidt's own, designed to work with the Superide components.

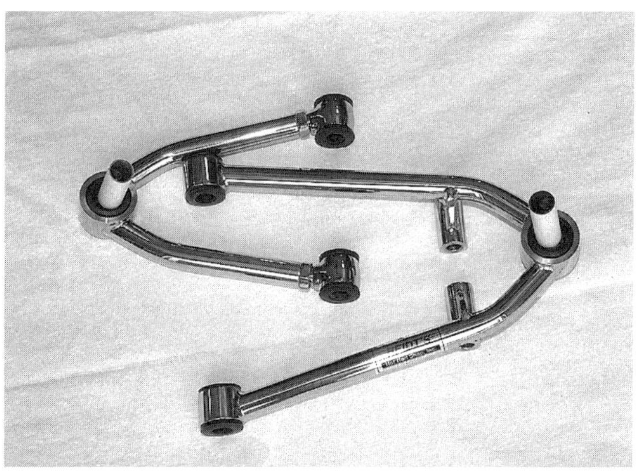

The upper and lower arms are built from polished stainless steel. Full-width lower arms eliminate the need for the strut rod and support bracket.

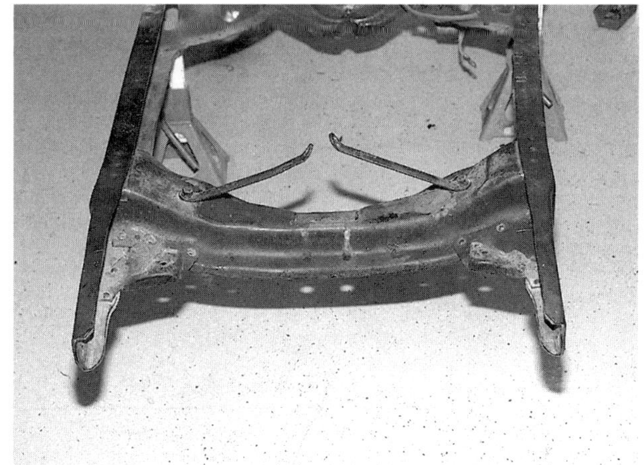

While working on the frame and before installation of the new cross-member it's a good idea to inspect the frame for cracks, rust and damage.

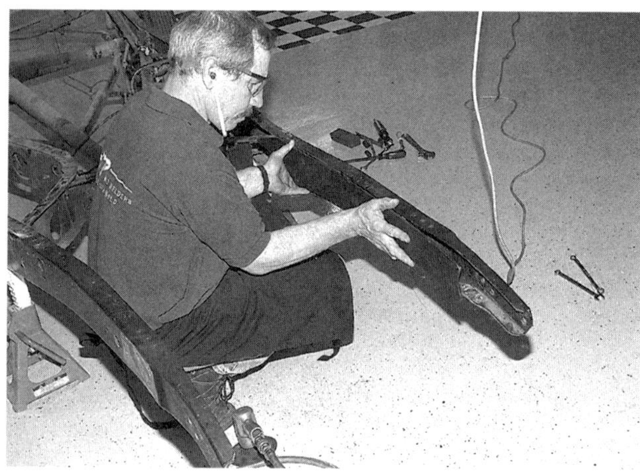

Satisfied that the original frame is in good condition, Jerry from HTP gets ready to install the boxing plates in the front frame rails.

Following some minor trimming of the boxing plates, Jerry tack welds them into position. Final welding is done slowly to avoid too much concentration of heat.

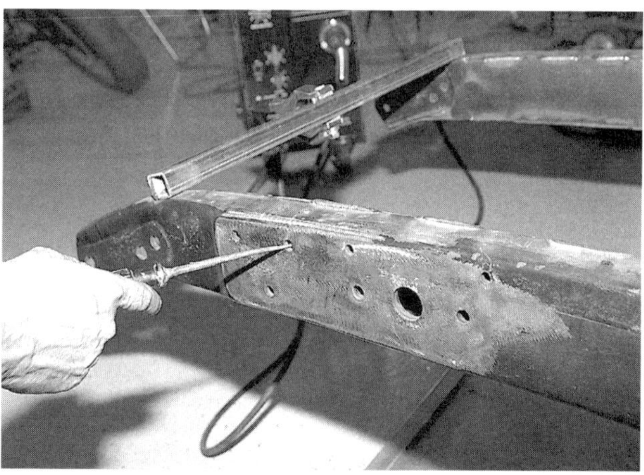

Note the tube track welded across the front of the frame to prevent the rails from moving. The second hole back from the front is a key measuring point for axle placement.

Point the Way

Three typical styles of steering linkage are used in most street rod applications: cross-steer, drag link, and rack and pinion.

Cross-steer systems typically mount the steering gear on the left frame rail with a long drag link that runs to the steering arm on the right side. Cross-steer systems are generally simple and easy to install. In most applications the pitman arm and drag link are kept neatly out of sight. This system offers minimal bump-steer and good steering control. Clearance problems can occur if the starter is on the left side or the drag link is too close to the oil pan.

"Drag link" style steering utilizes a long drag link running on the outside of the left frame rail, as seen on some older rods and many Bucket-Ts. The system avoids most of the clearance problems seen in cross-steer systems. The down side includes a tendency to bump-steer and a need to be very cautious with the specific layout for the sake of correct geometry. (See the illustrations.)

A rack and pinion system has the advantage of precision steering, good road feel and ready availability. Properly installed a rack and pinion system has a reputation for zero bump and roll steer. If the front suspension is an independent Mustang system or a Mustang clone, a Mustang rack and pinion is the only logical answer. The proverbial downside includes two major obstacles: First, the near impossibility of using a rack and pinion with a straight axle. Second, the difficulty in correctly mounting a rack and pinion to any front suspension unless that suspension was designed for the specific rack and pinion gear.

Bump steer

The term bump steer refers to the tendency of some suspension systems to effectively change direction, or steer themselves, when the car encounters a bump. Disconcerting to say the least, this little problem tends to occur because of a geometry problem. The problem is best illustrated by looking at a drag link system. Bump-steer in this system occurs when the bump causes the axle to move forward or back more than the drag link. The effect is a little different depending on the style of axle mounting but the end result is the same. Ideally the axle and the end of the drag link that attaches to the axle, must move through the same arcs as the suspension moves up and down.

In a cross-steer system the actual problem is axle movement to the side rather than to the front or rear. The gear should be mounted so the pitman arm is pointing straight ahead when the gear is in the center of its movement. It is also important that the gear be mounted so the drag link is parallel to the tie rod.

With a cross-steer system any axle movement to the side during suspension travel will effectively change the

The new cross-member is clamped in place and positioned so as to put the axle centerline in the correct position (which in some cases varies from stock).

length of the drag link and create bump-steer. As mentioned, street rods with a leaf spring, mounted buggy-style with a shackle at either end have the capacity to allow lateral axle movement on the shackles. Most street rod equipment manufacturers recommend a panhard rod in buggy sprung front axle installations, even though they are seldom seen on the street. Length and installation is critical to ensure that the panhard rod doesn't create more problems than it solves. If in doubt as to the need for a panhard or anti-roll bar consult the front axle manufacturer.

A builder should also consider the steering ratio and the effect that the pitman arm length has on that ratio. A longer pitman arm will provide faster steering - a given amount of steering wheel movement will result in more drag link movement. With manual steering and a big V-8 a very fast ratio will make for hard steering. Pitman arms are available for the most popular gears in different shapes and lengths which allow for fine tuning of the turning ratio, geometry and final gear position.

A cautious builder would be wise to mount the steering gear and linkage in a non-permanent fashion. At least until you are sure the steering geometry is correct and there are no clearance problems as the axle moves up and down and the wheels go lock to lock.

The drag link running from the gear to the steering arm must have an adjustment for length in order to get the steering wheel centered correctly while keeping the gear in the center of its movement. The right-side steering arm with double female ends necessary to attach both the drag link and the tie rod is available for most popular spindles. When you choose steering arms and tie rod ends make sure the tapered end of tie rod end is exactly the same size and angle as the taper in the steering arms. A slight mis-match in the tapers might have the end drawn up too far into the steering arm or loosening up later.

The Ackerman Effect, or "toe out on turns," is created by the shape of the steering arms (see the illustration). Heating and bending the steering arms

Based on the instructions, Jerry carefully marks the frame to ensure the cross-member goes in straight and that the centerline is located correctly.

The finished installation, complete with the steering gear and all the suspension components bolted in place. Of note, the axle centerline can vary on Ford frames, Heidt's suggests that you mark the original centerline or your own ideal wheel location as it is positioned in the fender well.

to clear a wishbone isn't a good idea and should only be done by an experienced shop. Try to buy the right combination of parts during the planning and mock up stage so no "fine adjustments" are needed later.

Steering gears

Vega steering gears, both new and used, are available from a number of sources. This has become the manual gear of choice for many modern street rod cross-steer installations. If you buy a used gear, consider the adjustment. If in doubt, most *Motors* and *Chilton* manuals have a section on adjusting one of these gears.

Some builders prefer a later GM manual gear, often called the 525. It is just a little larger and is often used instead of the Vega gear, especially in larger cars. Use of power steering on a straight axle street rod can be a problem due to the extra room the power gear takes up on the left side of the engine. The good news is the fact that the mounting plates available for GM steering gears will often fit either a manual or a power steering gear.

If you choose to install a complete Mustang II type suspension, then the gear usually comes with the kit, and a factory power rack and pinion is an option. These Ford power racks can usually be connected to a late-model G.M. power steering pump though the pressure of the pump may be too high for the Ford rack. In these situations the pressure can be reduced with a shim kit installed in the flow-control valve, or an adjustable power steering valve and hose kit from Heidt's can be installed.

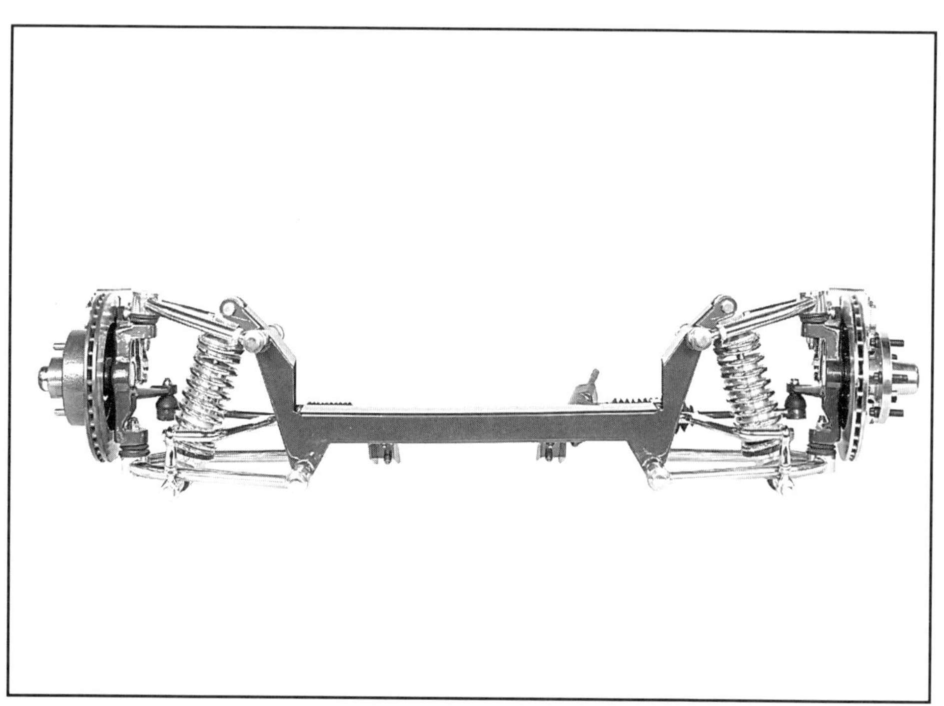

Various front suspension kits and assemblies are available for your new hot rod. The deluxe end of the spectrum offers independent suspension systems like this Superide designed from scratch to offer good handling and ride, and great aesthetics. Heidt's

Mount the gear

The general guidelines for mounting the steering gear include the necessity for rigidness. You want all the steering wheel motion translated into wheel motion. If the frame or

Alignment Basics

Caster, Camber and Toe

Understanding the basics of wheel alignment means first understanding a few specific terms. To help understand the concepts we've provided a few illustrations.

Camber:

Camber is the inward or outward tilt of the wheel at the top. Measured in degrees, camber is the tilt of the wheel from true vertical. If the wheel tilts out at the top the camber is said to be positive, a wheel that tilts in toward the car at the top has negative camber. The idea is to keep the tire planted flat on the road for maximum traction and grip. Sometimes a car manufacturer will specify something other than zero degrees camber while at rest in order to achieve good grip and tire wear while the car is going down the road.

Caster:

Caster is the backward or forward tilt of the spindle support arm at the top. Caster is a directional control angle measured in degrees, it is the amount the centerline of the spindle support arm (or king pin) is tilted from true vertical. The best example of caster is the front wheel of a bicycle. The fork tilts back at the top providing quite a bit of positive caster and a great deal of directional stability to the bike. If the fork were vertical (zero caster), riding "no hands" would be almost impossible.

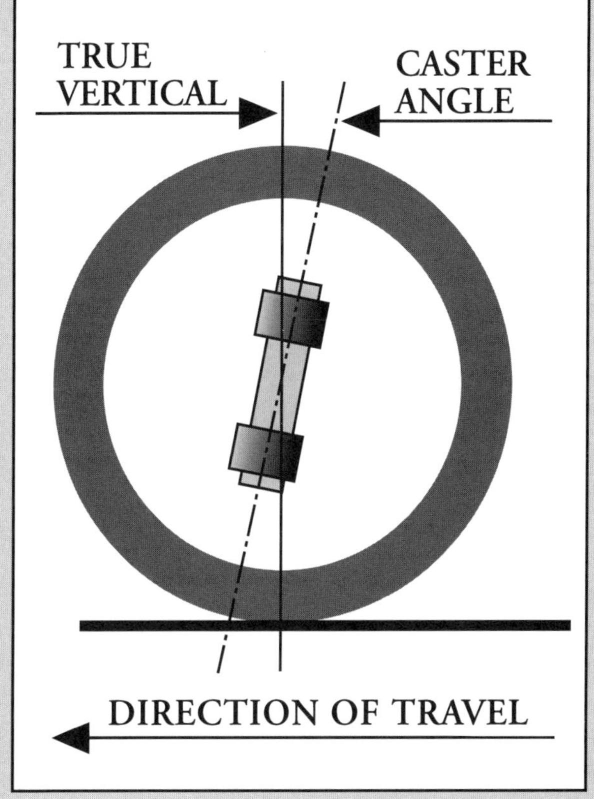

Caster is simply the tilt of the kingpin (or centerline of the ball joints) forward or back. Shown is positive caster.

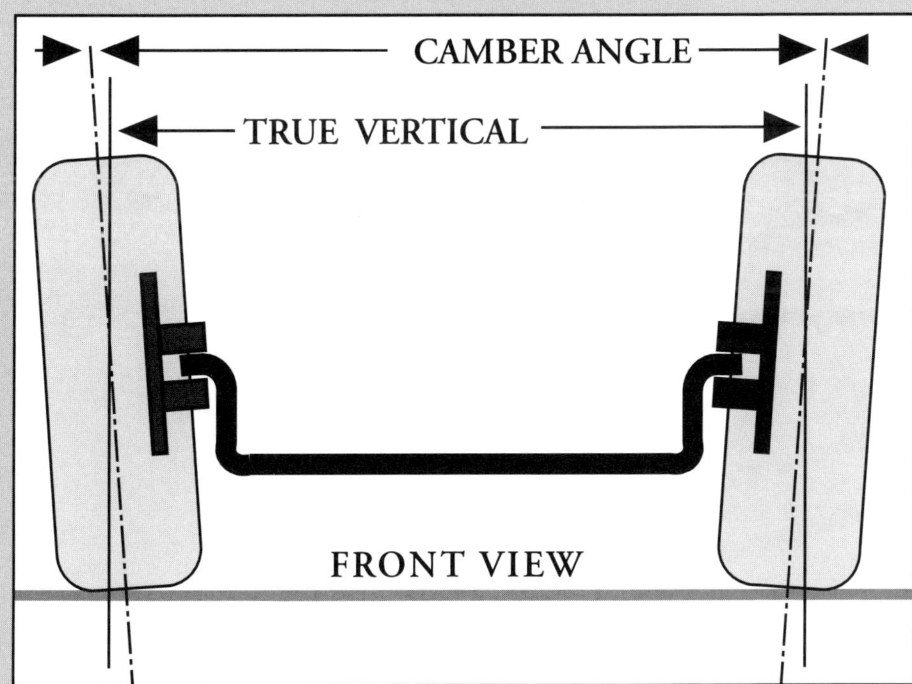

Camber is the tilt of the wheel from true vertical. A tilt to the outside is positive. With a straight axle the camber angle is determined by the axle itself. On independent systems the camber is nearly always adjustable.

Alignment Basics

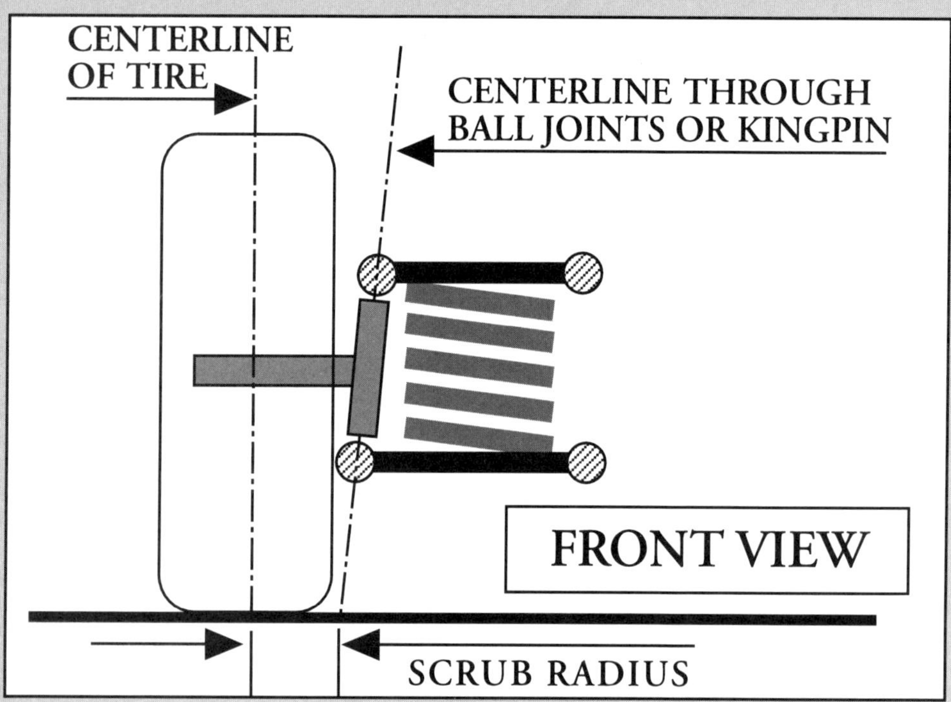

The distance from the centerline of the tire to that of the ball joints or kingpin is known as scrub radius. As the scrub radius increases the tire has more "leverage" on the rest of the steering linkage. It's best to use wheels with an offset close to stock.

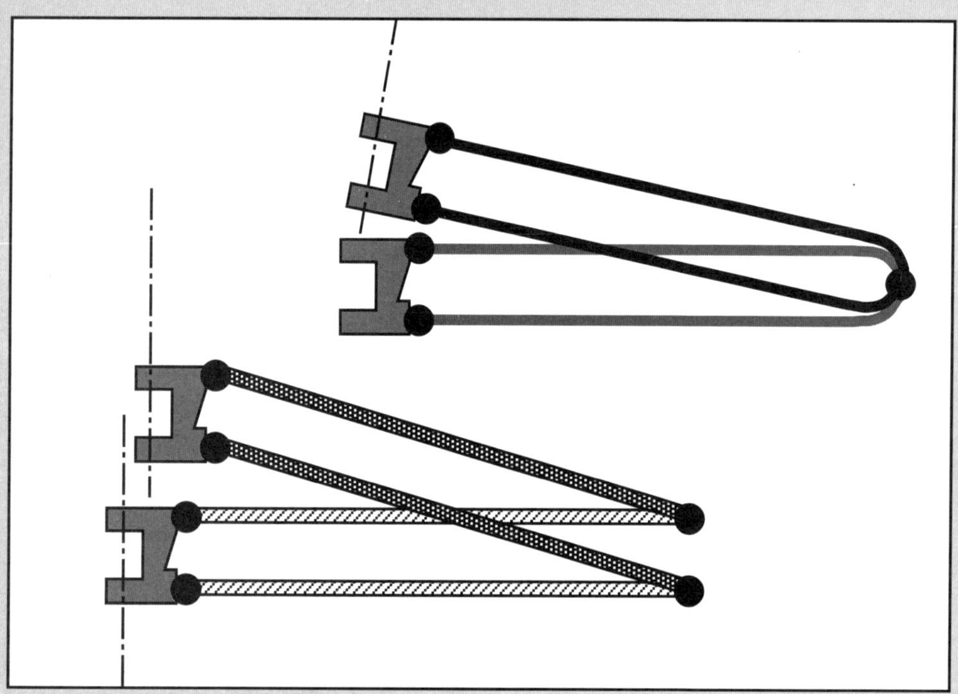

Radius rods and four-bar links have different effects on the axle as it goes over a bump. A four-bar linkage maintains a constant caster angle while radius rods of any kind cause a change in the caster angle as the axle moves through its travel.

Toe-In:

Toe-In or Toe-Out is the difference between a measurement taken across the front of the tires and one taken at the same spot on the back of the tires. The actual toe-in is seldom zero. A setting of 1/16 or 1/32inch is common - and done so that when the car is rolling down I-90 the load on all the steering and suspension pivots puts the two wheels parallel to one another.

And as long as we've gotten this far into the discussion of front suspension geometry, there are just a few more things to consider...

Wheel Offset:

Not an alignment angle at all but a very important part of the overall handling and suspension picture nonetheless. If you draw a line through the kingpin or the ball joints (front view) and then another line though the center of the tire, the distance between those two lines (where they intersect the ground) is called offset or scrub radius.

No matter what the actual measurement is, when you change the wheel and wheel offset on the front of your car you

Alignment Basics

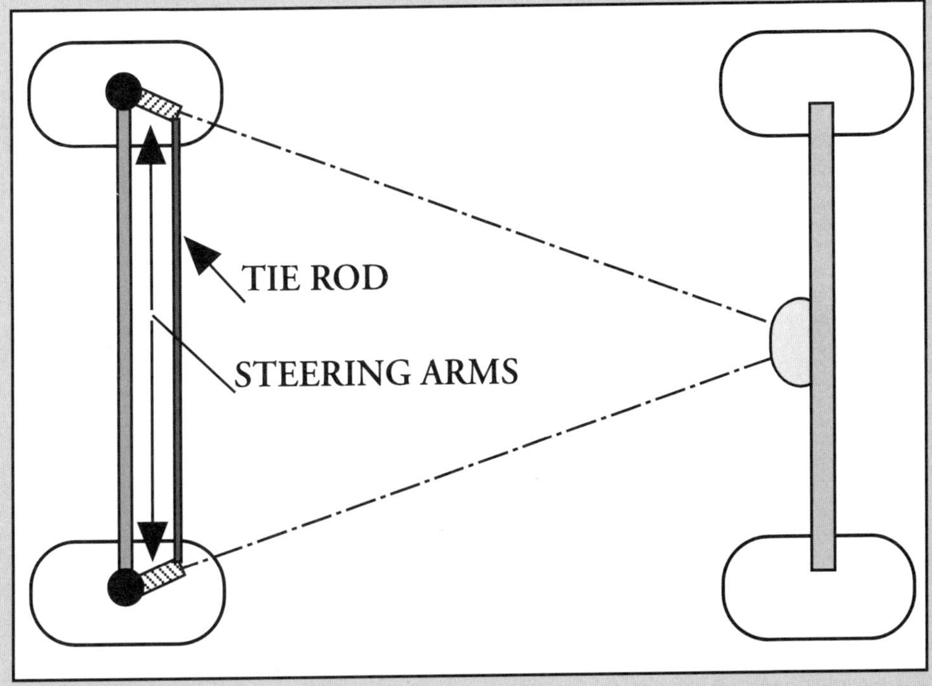

The angle of the steering arms is what creates the effect known as ackerman or "toe-out-on-turns." It helps to visualize the car from above.

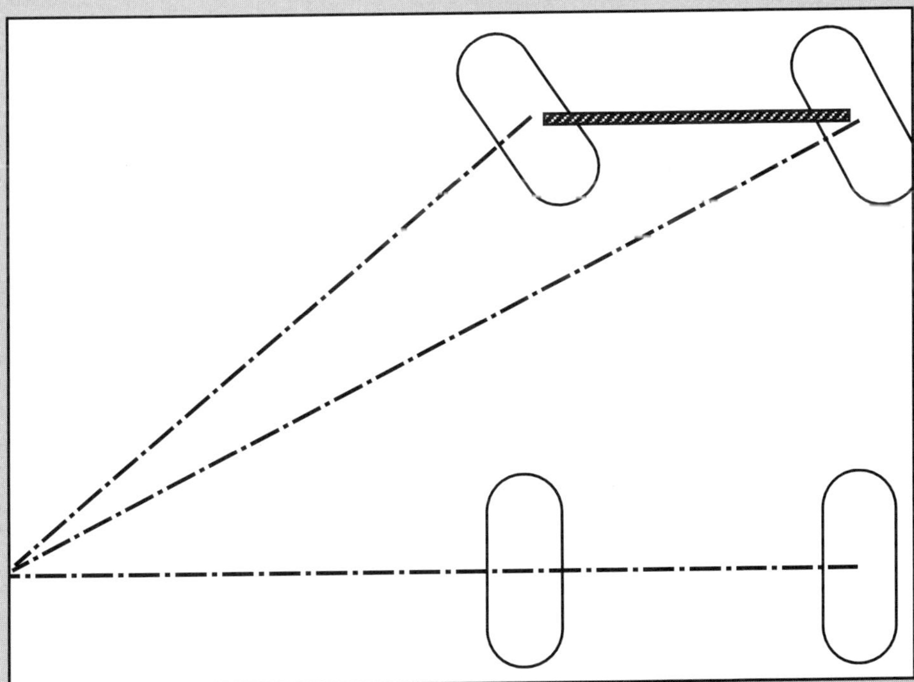

In a turn all the tires must rotate around a common center point. Because of the angle of the steering arms the inside tire always turns in a little more sharply. Typically, if the outside tire turns 20 degrees the inside turns 22 or 23 degrees.

effectively change the scrub radius. More wheel offset means more scrub radius and also means that now that wheel has a lot more "leverage" to act on the rest of the steering linkage. Changes in wheel offset from what the engineers intended will have profound effects on vehicle stability, brake performance on uneven surfaces, and "road feel."

Ackerman Effect:

Ackerman effect is the degree by which the inner front tire turns more sharply than the outer tire in a turn. In order to keep the car turning around a common center (and keeping in mind that the inner tire runs on a smaller diameter circle) the inner tire must turn more sharply than the outer tire. If the outer tire is turning twenty degrees from straight ahead, the inner tire usually turns twenty two or twenty-three degrees. (check the illustration again). This effect is essential to good handling and is created by the angle of the steering arms. The bottom line is the importance of correct steering arm shape and the sometimes frightening effects that come from rearranging the steering arm angles or location of the steering gear.

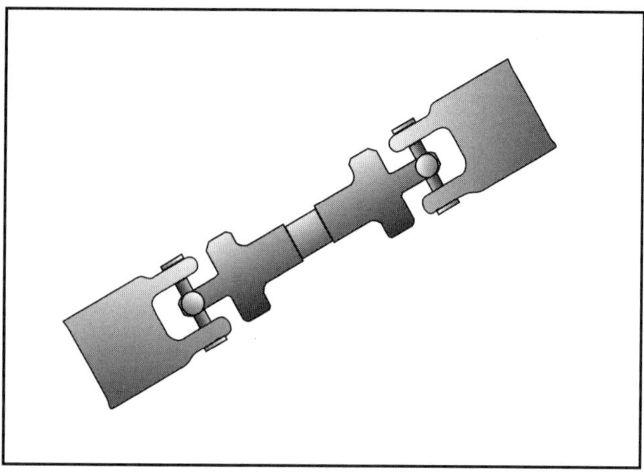

Whether the U-joints are part or a steering shaft or a drive shaft they must be "lined up" or kept in-phase.

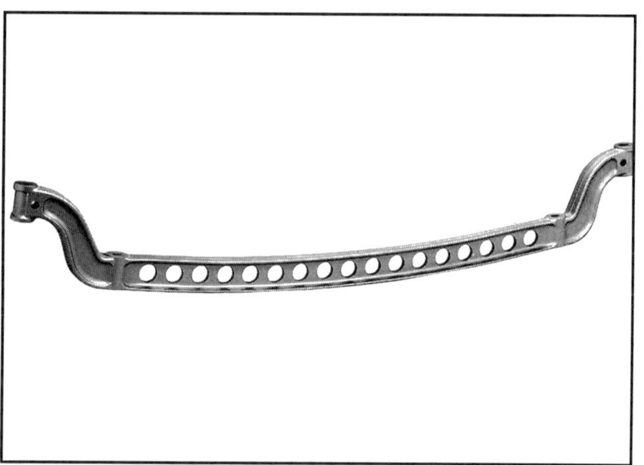

Available in 46 or 48 inch widths, these I-beam axles feature a four inch drop and come with or without ventilation. SO-CAL

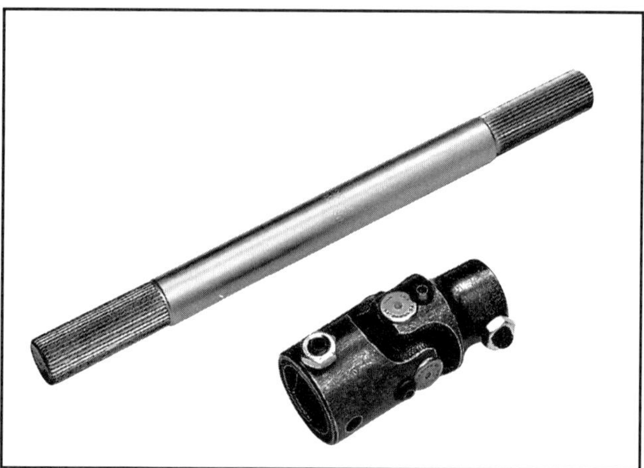

A variety of shafts and needle-bearing U-joints are available from Borgeson to fit nearly any application or situation. SO-CAL

steering mount flexes that flex will be transmitted to you as lost movement and sloppy steering. Just as important as rigid-ness is the necessity for correct geometry.

You probably want to clamp the gear mounting bracket (available from a variety of sources) in place, then mock-up your linkage and possibly the column as well, before doing any final welding. When you're mocking-up the gear and linkage, put the gear in the exact center of its movement, point the wheels straight ahead and center the steering wheel. This will put the gear in the center of its travel when you're going straight down the highway. An adjustable drag link allows you to precisely center the steering wheel with the gear in the center of its travel.

The gear needs to be in the center of its travel for a number of reasons, the biggest being that most gears have a "high-spot" in the center intended to compensate for wear that might occur in the straight ahead position.

Street rod builders often mount the steering gear so the worm gear is parallel to the frame rail. This is done so the pitman shaft ends up perpendicular to the ground and the pitman arm moves back and forth without going up and down. This set up may make it easier for the drag link to clear the oil pan, yet it also makes for a more complicated shaft from the column to the gear. You don't have to mount the gear so the worm shaft is parallel to the frame rail. New cars are often built with the worm gear pointing at the column in order to avoid a complicated shaft with three U-joints.

No matter how you mount the gear, be sure there's sufficient room between the oil pan and the drag link when the suspension is bottomed out and try to keep the drag link parallel to the tie rod.

The steering-gear mounting plate should be mounted to a boxed frame rail. When you finally bolt that gear to the mounting plate, use grade eight bolts and a little Loctite. As stated before, any movement in the mounting plate will translate into sloppy, vague steering.

Note: A recent tech article in *Street Rodder* points out the fact that installation of some aftermarket pitman arms allows the retaining nut to bottom out on the threads before the pitman arm is actually pushed tight into place. If in doubt use a large washer to ensure the pitman arm is tightened securely on the shaft.

A panhard rod should be kept parallel to the drag link and be the same length (or as close as possible) as the drag link - so it moves through the same arcs as the drag link when the suspension moves up and down.

Before you do the final welding of the brackets, make one final check for clearance of all the steering linkage parts with the suspension in both extremes of travel. Starting in the center of steering-gear travel with the wheels pointing straight ahead, make sure that full lock is the same number of turns in each direction.

Mounting the column and shaft

In choosing a column many street rodders buy a late model GM unit. While certainly not the only game in town, most GM columns have some kind of slip fit, extendible shaft. This means that adjustments in length can be accomplished by disassembly, shortening or lengthening the center shaft and then sectioning the outer case. This is one of those details you should keep in mind if you can't find just the right column.

When shopping for columns consider that most late-model columns are designed to collapse in an accident. A nice feature even though we all know it can't happen to us.

There are a few basic rules for mounting the column. First, the column should be mounted to provide a good angle between the column shaft and the steering gear. Second, and often overlooked, the column should be mounted to provide a comfortable driving position. With the possible exception of Bucket-Ts, the lower column should pass through the firewall high enough that it doesn't interfere with the placement of your feet or moving your foot from the gas to the brake pedal.

Third, the mounts used to attach the column don't have to be quarter-inch boiler plate. Plate in the 3/32 to 3/16 inch range is certainly adequate. Fourth, the column must be mounted at both ends, which means some dash boards may need reinforcement. If there is a case for overkill on the mount it's on the lower, not the upper, column mount. Finally, it's a good idea to mount the column in a temporary fashion and then double check all the angles and clearances before having the new brackets fabricated.

A flexible coupling is needed between the column and the steering gear. Detroit usually uses one or two U-joints and a "rag" joint. For street-rod use Borgeson makes a damper. In either case, the flexible joint helps to isolate road noise and compensate for any shifting between the body and the frame.

U-joints used as part of the steering shaft must be needle-bearing joints. Those without needle bearings are often available, but are intended for industrial or racing applications. These high quality U-Joints should be kept in their working range, usually less than thirty degrees. When more than one joint is used they must be kept in phase and an additional support bearing must generally be added.

Always remember that the best systems are the simple ones and that while the steering gear must be mounted for correct geometry there is usually some flexibility in the final position. Don't go to a lot of work designing a complicated shaft with three joints just to get around an exhaust pipe, which can probably be moved or modified with relative ease.

Trouble occurs when the wrong type ends are used on the four-bars. Do not use heim joints or spherical rod ends. Use the style shown here or buy a quality kit. Chassis Engineering

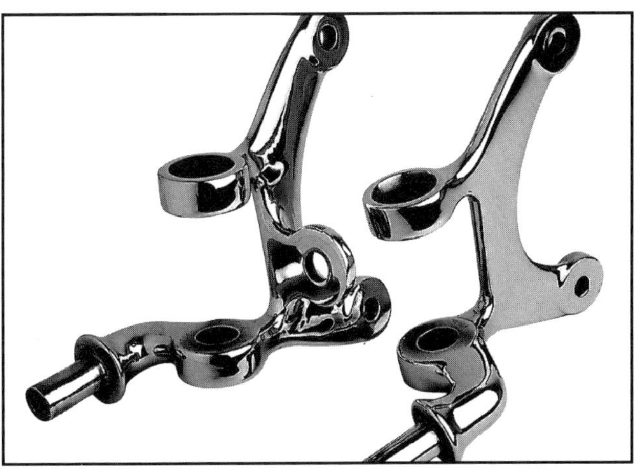

These new polished stainless batwings feature integral lower shock mounts and a Panhard bar mount. SO-CAL

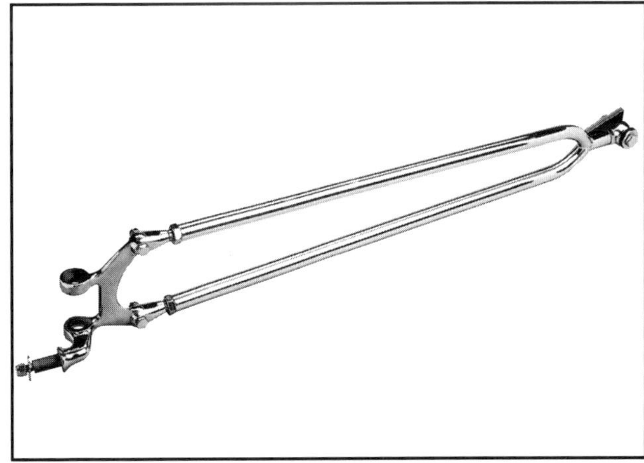

If yours is a nostalgia car you'll want to use hairpin radius rods to locate that I-beam axle (tube axles need a four-bar linkage). SO-CAL

Fat Man offers an upgrade kit that allows owners of Mustang II suspension kits to install air suspension in place of conventional coils. Fat Man

Q&A: Pete Chapouris

Q: Pete, I think everyone knows your name, but some people might not know much about the new shop. Can you give us a quick rundown on the new SO-CAL shop and the kind of work you're doing.

A: Well, I did a semi-retirement thing up on the mountain for a few years, and then my wife and I decided to move back into the valley. I didn't want to move into a two car garage and start over real small, so I gathered up some workers and started the Pete Chapouris Group. That went OK but I realized it was going to take a long time to establish my name. At the same time I was working with Alex Xydias and Bruce Meyer on the restoration of the SO-CAL Belly Tank. Later I decided to ask Alex if he would license the name of the original SO-CAL Speed Shop. Basically, he said yes, so we're off and running with the new company.

Currently we have three retail stores that feature our own products as well as parts from 15 companies that we represent, companies like Edelbrock, B&M, Pete & Jakes and Hedman. We also have a large shop with 11 craftsmen. The foreman is Shane Weatherly from Arizona, he's a young man with all the right ingredients to be a big name in this industry. There are 22 cars on the floor now but eventually I'd like to narrow the focus to only six or eight per year. The stores are important, but the shop is heart and soul of the company. We will always have the shop. We're planning to take 5 cars to Oakland for the Fiftieth Anniversary and we're working with both *Rod&Custom* and *Street Rodder* on magazine project cars.

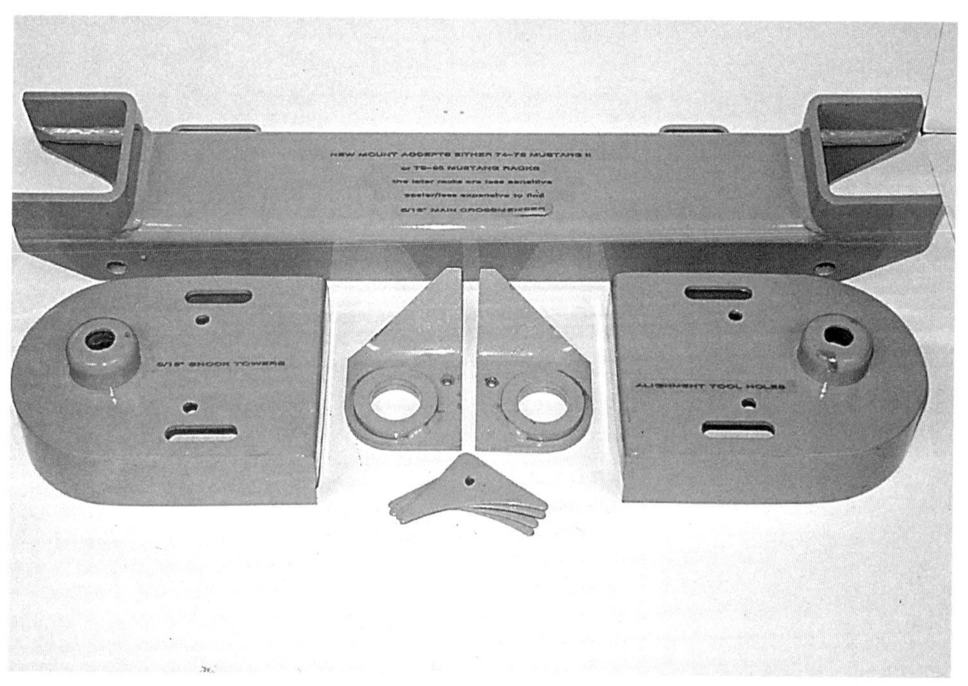

Cross-members designed to accept Mustang II components are available to fit a wide variety of chassis, including the '40 - '48 Chevy kit seen here. Fat Man

Q: Let's talk about front suspension, if I come into your shop to discuss a new car you're building for me, how do we decide on the type of front suspension that's right for that car. What kind of parameters do we discuss?

A: For me if it's an open wheel car, it should have a dropped axle. With an old car when you take the axle out of it you've lost all the charm, you loose everything. If it's a fat car then we might want to put in an independent suspension. We've built four cars recently where we put in modified Mustang II systems. You can't see the suspension and it's a different kind of car with more creature comforts. The style of suspension should be dictated by the car.

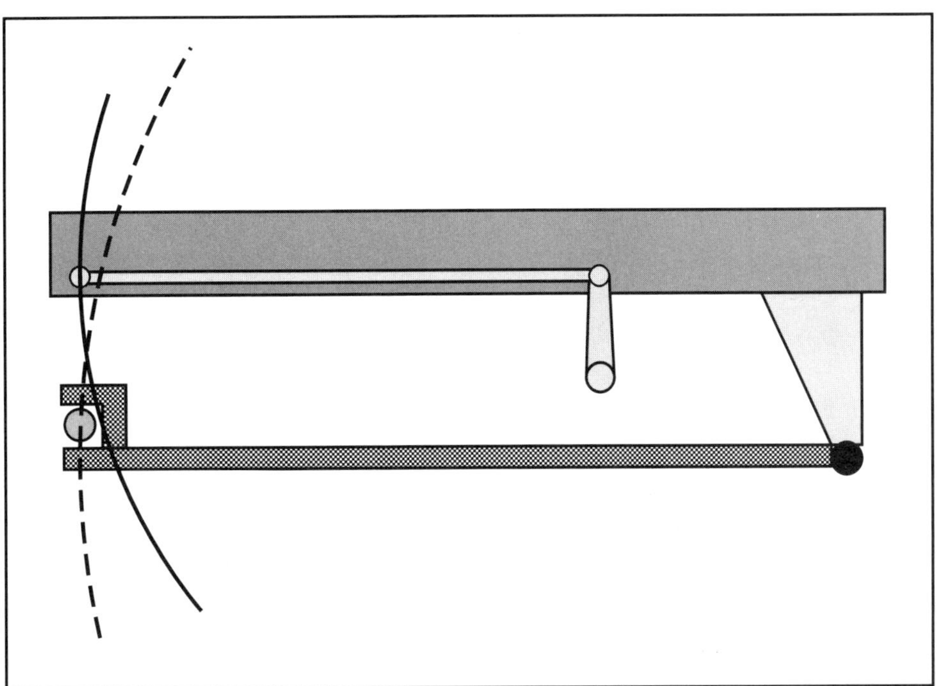

This diagram shows how, with a drag-link steering set up that's poorly engineered, the axle and steering linkage move through very different arcs during suspension movement. "Bump steer" can plague a cross-steer setup too. Use a known kit or design carefully.

Q: If we agree that the Deuce hiboy should have a dropped axle, how do we choose between four-bar linkage and a split wishbone?

A: Basically it's a time-frame deal. In the sequence of events, when Super Bell came out with the tube axle we (Pete & Jakes) developed the four bar linkage. (Note: you can't use a tube axle with a split wishbone because the system puts a twist on the axle which a tube axle won't tolerate.) Then later we started using dropped I-beam axles and we still put on a four bar system. But the traditional look is a hairpin. If it's a deuce you have to ask yourself, should that car have a four-bar? The industry offers both and that's what's neat. That our industry has evolved to the point where you can have what you want, depending on your personal taste and the aesthetic considerations.

This Mustang II cross-member and suspension uses air bags in place of coil springs to provide adjustable ride and ride height. TCI

Q: Do you prefer a cross-steer linkage, and which steering box do you typically use.

A: We always set them up with a cross-steer linkage, unless it's some kind of restoration on an older hot rod. And we always

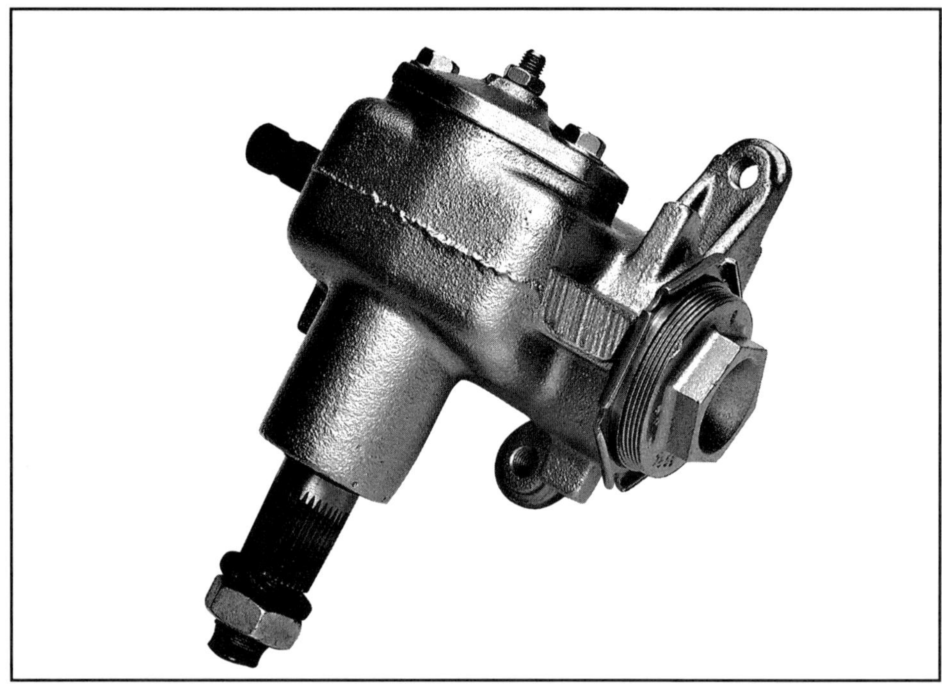

These Mullins/Vega steering boxes are brand new, built by the original manufacturer to the original dimensions. SO-CAL

use the Mullins/Vega steering box.

Q: Do you prefer to see straight axle/buggy spring installations use a panhard rod to prevent bump steer?

A: If the car has exactly the right angle on the shackles you can get along without a panhard rod, but as the car gets older you're likely to run into trouble. Basically, if you use a buggy spring and a cross-steer set up then you need a panhard rod. We also like to install a steering damper, it ensures there is no side shimmy. Our straight-axle installations always include a panhard rod and a damper.

Q: With a straight axle, what do you like to see for wheel alignment settings.

A: We always try for zero camber, and a minimum of 5 degrees of positive caster. Actually, 5 to 9 degrees of caster. My new car runs 7-1/2 degrees. We set the toe in at 1/16 to 1/8 inch though sometimes if the front wheels have a lot or offset to the outside, which moves the wheel away from the kingpin, then you might need a little bit more. The camber is determined by the axle of course and the Super Bell and Magnum people have it figured out so when the weight goes on the car the camber comes out to zero or just a little bit positive.

If you're looking for something more exotic than the mundane Mustang suspension, consider kits designed to utilize Corvette suspension and brake components with coil-over springs. Fat Man

Q: Can you talk about the front cross-member in your new catalog. Can you explain the importance of the cross-member and the relationship between it and the caster angle?

A: The cross-member is manufactured by TCI, but we make sure the center part of the cross-member is dialed in when you get it.

Most people forget that the car is on a rake of 3 to 5 degrees. So if you've got a cross-member that's flat, when you rake the car the caster goes negative. You need nine degrees of positive caster to net out at six. People build cars flat in the garage and then after they put the rake on the car the caster is all wrong. So our cross-member allows you to run more positive caster without putting a bind in the spring or spring shackle. This cross member also lowers the front end of the car one inch.

Q: What are the mistakes people make in choosing and installing a front suspension.

A: What we see most is the wrong front spring, it's too short or too long and then they have the wrong shackle angle. Sometimes they take out spring leaves to lower the car and that doesn't work either. They use the wrong front cross-member which we talked about. They select the wrong wheels with too much offset. People don't realize the importance of choosing the right wheels and tires.

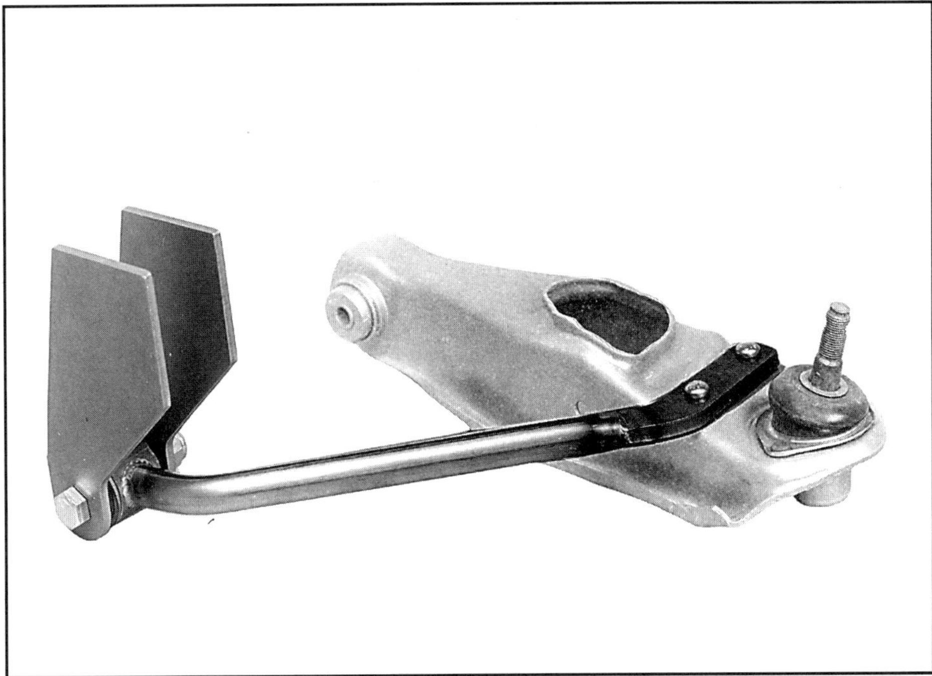

Lower control arms for Mustang-type suspension kits come two ways. Shown is the narrow based lower arm with strut rod, as opposed to the full-width lower arm that do not use the strut rod. Heidt's

Q&A: Gary Heidt

Q: Gary, can we start with a little background on you and Heidt's, how you got started making components for street rods and how the company has grown since you started?

A: I started out as a street rodder, I went to the Nationals in 1974 without a car, then I bought a 1/2 ton T-bucket which I fixed up and drove to the Nats in 1975. Since 1974 I've been to every National. After the T-bucket I built two more cars and I was doing work on friends cars and fabricating some parts in my home shop. The third car was a '33 Ford coupe and I installed a Jaguar front suspension and independent rear suspension. About that time a friend of mine who ran a street rod shop asked me if I'd ever considered making Mustang II suspension kits for hot rods. That was the fall of 1985. I made the first one by hand and we installed it on a customer car

Here you see a typical "split wishbone" on a frame under construction at Chassis Engineering.

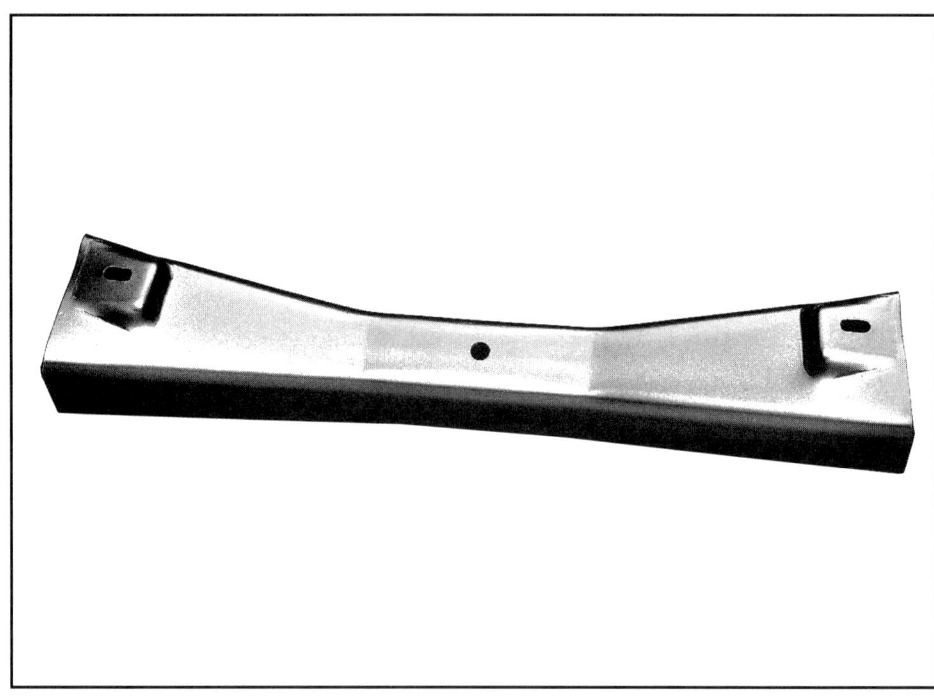

Not all early Ford front cross-members are the same. Pick one that provides enough caster (especially if the car will sit on a rake) and that works with the rest of the front end components to provide the ride height you're after.

in my friend's shop. We took pictures of the installation and street rodder ran the story. I took out a small ad at the same time and it took off like a rocket. That was April of 1986. That's when I quit my job as a design engineer and worked full time on Heidt's Hot Rod Shop. Since then we've added more and more parts and kits to our catalog. The thing I'm most proud of is the acceptance of our Heidt's parts by the industry.

Q: Gary, a lot of cars can be set up to run either a straight axle or an independent suspension. How do you advise people who call and aren't sure which type of axle to run?

A: I ask what type of car is it, modern or nostalgia. An independent rides better, but if the car has to have a certain look then an independent doesn't do the job. If it's a high tech car it should be independent, if it's a traditional car then a straight axle is probably a better choice.

Q: Once a person decides to run an independent system how can they tell which of the kits on the market is the best one, in terms of good handling and ease of installation?

A: There are all kinds of kits, good and bad. The way I describe a good kit is, 'It will only fit a particular application. It's not a universal kit.' It's like selling a wheel with no lug nut holes and then you drill them out to the right pattern. With universal kits, if you don't cut the parts in the kit properly it will never fit right. You should only have to place the parts not cut them. The universal kits are hard to install properly. The good reputable kits almost install themselves.

(Note, by universal we do not mean examples like kits made for 1935-1940 Fords. In this case one kits fits all the years because the frames are the same.)

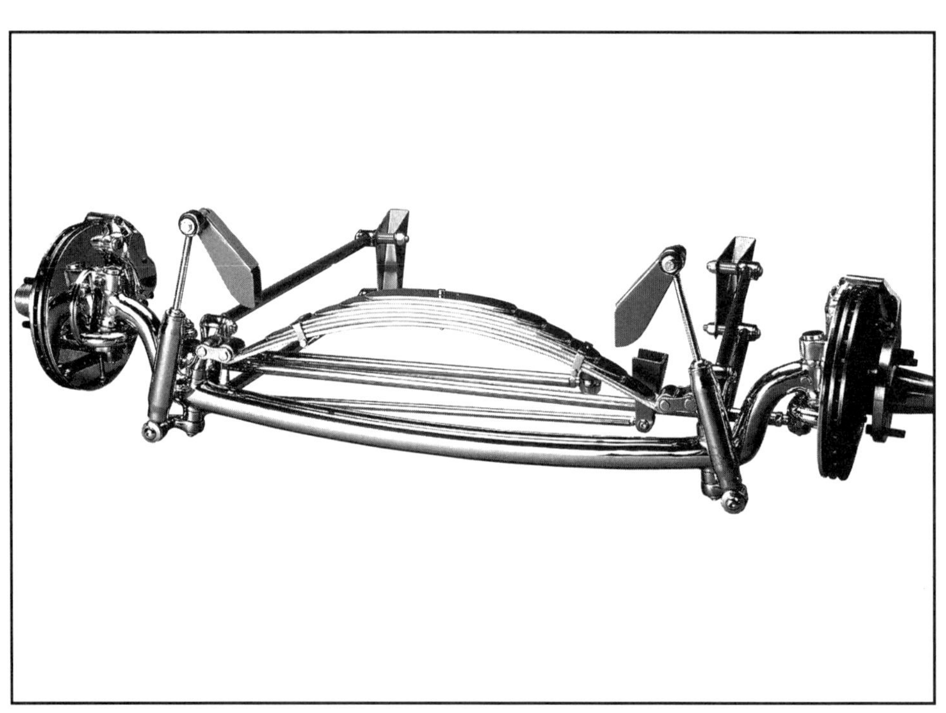

This chrome plated tube axle uses a chrome spring and stainless hardware. If your budget is tight, similar assemblies are available in paint. TCI

Sometimes it helps to ask the manufacturer for the installation instructions, so you can see exactly how much work it is to install the kit. We willingly give out our instructions. If they are afraid to send out their instructions then you may not want to install their kit.

In terms of handling, be sure the kit has anti-dive built in. The Pinto used three degrees and the Mustang had seven, We use three degrees of anti-dive and you should be able to see three degrees when you look at the installed A-arms on a finished frame. Our kits are designed to fit stock parts and use the stock Ford geometry.

Q: What if I have a car that no one makes a kit specifically for?

A: We do make universal kits for those situations, but they are labeled as such. They work for most cars, except some of the real heavy '50s cars like a '55 Cadillac.

Q: What are the advantages of installing a deluxe kit like the Superide instead of a simple Mustang II kit?

A: The Mustang II kits are the backbone of our product line and they are excellent products. The Superide, however, is deluxe in the sense that it comes standard with 11 inch rotors, tubular A arms and a full-width lower arm. The major difference is perceived value, increased value. Superide is a higher status product.

Q: What about the difference between kits that use the stock Pinto lower arm with the separate strut, and the more deluxe kits with a full-width lower A-frame?

A: They handle the same, but the full-width lower A-frame is a better design. Better because you don't have to weld on the supports for the strut rod. That's a difficult weld to do correctly, sometimes people do it from underneath. It's better to just eliminate the potential problem by using the full-width lower arm.

Q: Can people at home install these kits or should they be installed by a good street rod shop? Do inexperienced people get in trouble when they try to install these front end kits?

A: I talked to a guy a the SEMA show recently. He just installed one of our kits and said it fit really right. Our kits, and most good ones, are intended to be installed by the owner at home. We provide good instructions and good diagrams. The owner only needs to make one or two reference measurements. And the instruction sheets show the exact application.

Q: Do people sometimes buy the wrong kit or wrong brand of kit for their car and get in trouble that way?

A: There are problems with some of the universal kits we already talked about. Sometimes the parts don't fit right or the cross-member wasn't installed quite right so then the car can't be aligned. One thing to look for before buying a kit is good technical support. Is there someone on the other end of a phone line who can answer your questions or help you through a rough spot in the installation?

Q: What about alignment, do you provide specs, are these the same as the Mustang specs?

A: The alignment specifications are the same as the stock Mustang II.

Heidt's offers upgrade kits that convert existing Mustang II systems to air suspension, or a complete air suspension cross-member and suspension kit. Both come with their bottom out snubbers. Heidt's

Chapter Four

Rear Suspension

Solid or Independent?

Basic systems

Rear suspension systems come in a number of different styles. The possibilities here range from a simple solid-axle with buggy or leaf spring(s) to a fully independent system based on a Jaguar or Corvette rear end.

The cost, handling and style are the three things that determine which of the many possible types and brands will best match up to your project.

In the broadest sense there are two kinds of rear suspension: solid or independent. Nearly all the solid axle designs can be divided by the type of spring: leaf or coil. As mentioned, some leaf spring designs use two parallel

Hot rodders can argue the various ride and handling advantages of independent suspensionn all day long. But the real reason people spend the extra time and money to install independent rear suspension is much simpler than that. People like independent rear suspension 'cause it looks so darned cool. Kurt Senescall

leafs while others use the tried and true buggy spring.

Leaf springs are readily available and relatively inexpensive. They also distribute any weight bias on one side to the whole chassis and are capable of handling a lot of torque. The springs locate the axle as well as support the car, eliminating any need for a panhard bar. Depending on personal tastes, a leaf spring rear suspension provides reasonably good handling and ride. While not the most sophisticated of suspension systems, a simple set of leaf springs supporting a 9-inch Ford or 10-bolt GM rearend represents a good buy-for-the-buck.

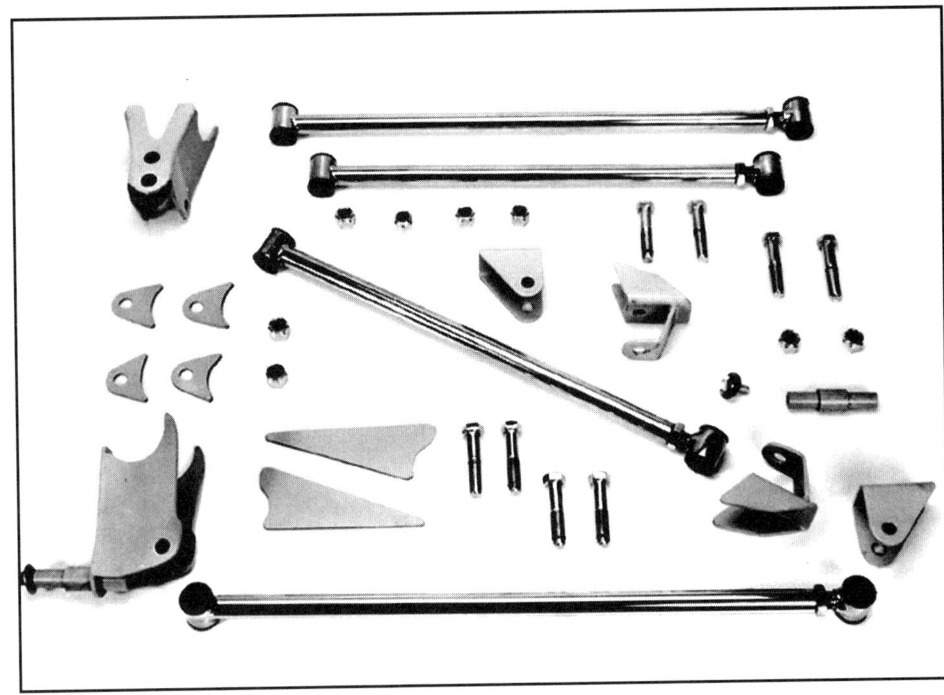

Triangulated four-bar systems eliminate the need for a Panhard bar and are available for hot rods of almost any description. Chassis Engineering

Leaf springs have the problem of "stiction," or friction between the leafs. The demerits column also includes a lack of adjustability and minimal visual appeal. Posies and some other firms make "Super Slide" springs with little Teflon buttons between the leafs to minimize friction. They also make single leaf springs for many applications. Teflon strips and buttons are also available to eliminate the inter-leaf friction. To enhance the visual appeal of your leaf springs, chrome assemblies are available from a number of sources or you can have yours chrome plated.

You can lower a leaf-spring suspension with lowering blocks. Pulling one or more leafs from the spring pack may not be a good idea as the spring pack was designed to work as an "assembly." A better solution would be to have the springs de-arched, a process that can be performed by most large spring companies.

Rear coil spring suspension systems come in many styles, much like systems seen on the front of our hot rods. Most coil spring systems allow for easier height adjustment due to the ease of swapping springs and the adjustment collar on most street rod coil-over assemblies. Coil

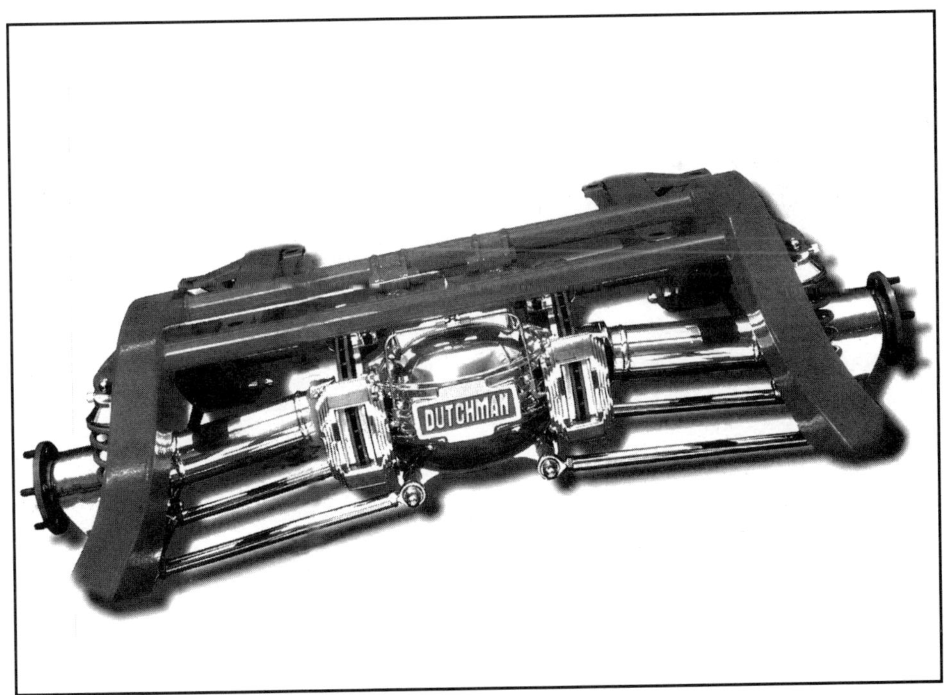

Among the non Corvette or Jaguar independent systems on the market is this complete assembly from Dutchman based on the venerable nine-inch Ford rearend. Features include inboard Wilwood brakes, 31-spline axles, and polishes axles. Dutchman

Seen in Jon Kosmoski's shop, this Mike Adams' tubular '34 Ford chassis supports a narrowed '84 to '87 Corvette suspension assembly. The chassis is built from 1-3/4 inch mild steel tubing designed to be both lighter and stronger than a typical boxed frame rail.

spring systems are generally considered to have good visual appeal and some of the designs do a good job of building traction if yours is a true hot rod project.

The downside includes the need for a panhard rod (in most systems) and the increased cost if you are converting an early rod from the original leaf springs to a coil spring system.

The four-bar rear suspension is one of the most popular of the many coil-spring suspension systems. These four-bar systems actually come in two versions: parallel and triangulated. Parallel systems run two pairs of parallel rods forward from the rearend housing to the brackets on the frame, much like the system used on the front of many street rods. In most applications the coil-overs mount behind the axle. The parallel rods locate the rearend housing fore and aft, allowing it to move up and down. A panhard rod is used to locate the rearend laterally. This rod absorbs side thrust and prevents excessive side-to-side movement of the rearend relative to the frame.

Four-bar systems have the advantage of being easily purchased from a number of street rod suppliers. Most kits come with coil-over shock units matched to the most popular cars. The system offers good ride, handling and easy adjustability.

Four-bar systems also come in a triangulated four-bar package. This system uses two rods running forward from the rearend housing to brackets on the side rails of the frame - these rods run parallel to the side rails. The other two rods run from the rearend housing to the frame at an angle relative to the frame rails. These two "angled rods" absorb side thrust and eliminate the need for a panhard rod. Some builders don't feel these systems handle

Mike Adams makes his own toe and camber rods, and uses narrowed factory axles. The system takes advantage of the Corvette's light weight components and good handling characteristics.

hard corners as well as a pure four-bar and complain further that it's hard to route the exhaust with the upper triangulated bars in the way.

There are also a few three-bar systems. Though seldom seen on street rods, the three-bar has been around a long time and is often used on drag race applications. A three-bar uses two long, parallel bars and one shorter, top bar. This systems uses a panhard rod to absorb side loads. The major advantage is simplicity and traction.

The simplest of the coil-spring systems is a set of ladder bars. Like a modified four-bar, the ladder bars mount to the rearend housing and to a common point on the side rail of the frame. Ladder bars are simple, adjustable and readily available. These bars transmit power directly to the frame, at a point well forward on the frame. Ladder bars are popular with drag racers as they help give the car a good launch when the light blinks green.

These systems do require a panhard rod and don't seem to offer the visual appeal of chrome four-bars. Because the bar mounts to a single front pivot there is also considerable change in pinion angle as the suspension moves up and down. The amount of pinion angle change depends on the length of the ladder bars, longer bars suffer less change. Some hot rodders feel that ladder bars don't handle in the corners as well as the other systems, and the long bars can get in the way of the exhaust routing.

Though generally more expensive than any other type of rear suspension, an independent rear suspension offers that unique combination of good looks and good handling. Besides money, the cost of a nice independent system includes additional complexity and possibly increased maintenance. The real good news in the independent rear suspension department is the number of kits currently available for many applications that make it easier than ever to put an independent suspension system under anything from a Model A to a fat-fendered Chevy sedan.

What to buy

Before you buy a leaf spring kit be sure the kit comes from a quality manufacturer and is truly designed for your car. Ask if the springs are de-arched and if there is any provision for height adjustment. A four-bar system may be more difficult to purchase simply because of all the different kits and designs.

If you build your own system be sure the bars are long enough to provide good geometry and that all the components and rod ends are strong enough for the task at hand. Any panhard rod should be as long as is practical so that as the suspension moves up and down the body will experience less side-to-side movement.

Ladder bar kits should be purchased with the same

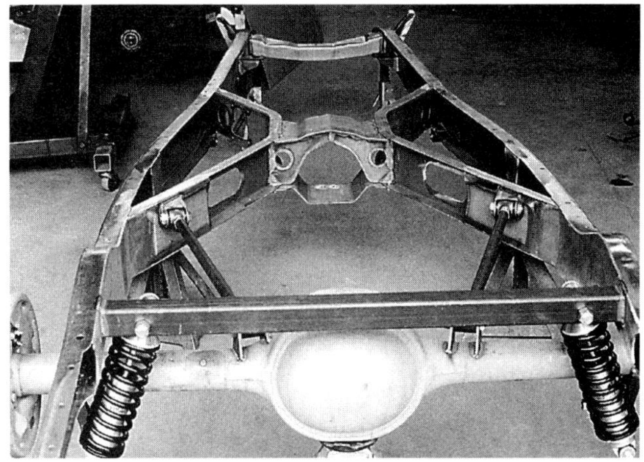

Built at Chassis Engineering, this frame uses an interesting traingulated four-bar system - the lower bars are the split wishbone. Eric Aurand

Detail shot shows the coil-over, the lower link and the fabricated upper link with brakcets added to the housing. Eric Aurand

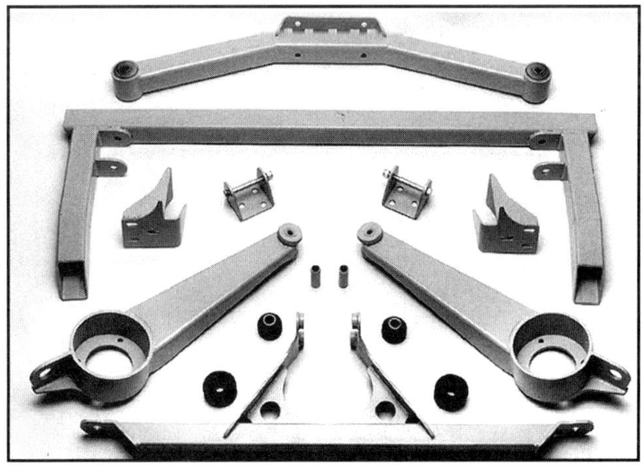

Adapt earlier Corvette independent rear suspension to your Model A with this kit that includes cross-member, trailing arms and shock brackets. Chassis Engineering

Riding on a Cushion of Air

They say you can't have your cake and eat it too. That blown big-blocks will never get twenty miles per gallon and that the really slammed cars will never be easy or practical to drive around town.

Times have changed and though big hairy engines still don't get great mileage (unless you shut off the key going downhill) you can now have a ride that truly sits "in the weeds" at the fairground yet will negotiate speed bumps and driveways on your way home from the show.

The answer to our prayers is the new air-suspension systems offered by companies like Art Morrison, Heidt's, Fat Man Fabrications and others.

The systems described here rely on air springs from Firestone or Goodyear, which take the place of the conventional coil or leaf spring typically used in most of our vehicles. Some of these systems have been designed from scratch around the air springs while others use coil-replacement therapy to put air bags where coils once lived, often in Mustang II suspension systems.

The air springs are connected through a control panel inside the car to a small electric compressor. Though this might sound sexy and high-tech, air springs aren't new. They've actually been around for decades, used primarily in over-the-road trucks. Air springs have at least one characteristic that makes them such a good starting point for suspension systems. Unlike a coil spring which has the same rate at one inch of compression as it does at three, an air spring progressively increases its spring rate as it is compressed.

Art Morrison describes the Firestone air springs used by his company as very tire-like in their construction. "These are built mostly by hand, much like a tire, at the big Firestone plant in Indiana. The body is two-ply construction and the rubber is bonded to the cords with heat and pressure, it's a vulcanizing process. The air springs used on street rod suspensions are made from the same very tough, two-ply material used to build all the other air springs."

In terms of durability, the manufacturers commonly test their air springs into the millions of cycles without failure. Or what a Firestone engineer described as, "the equivalent of about 40 years of continual use on a truck."

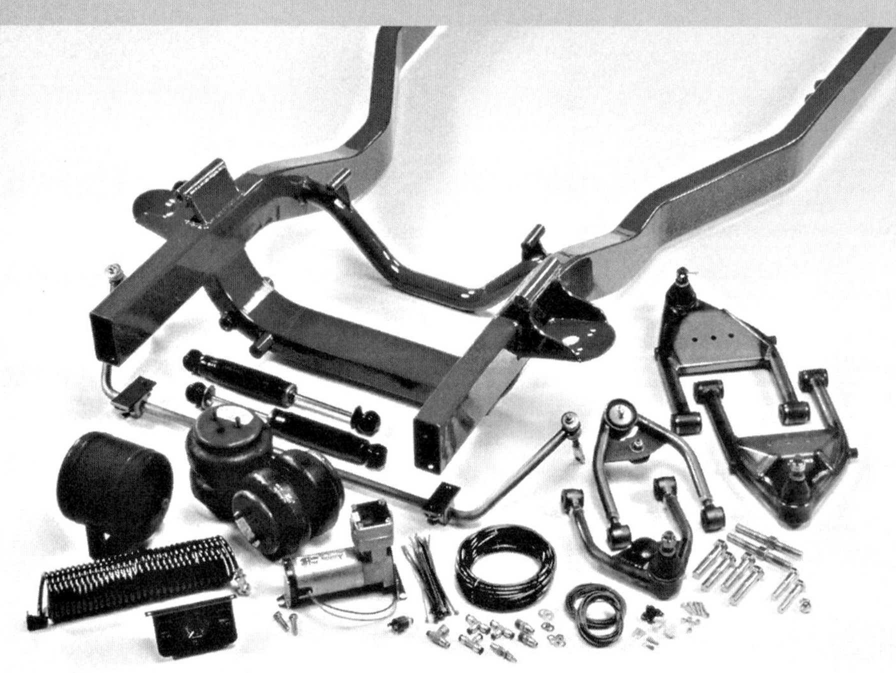

Air suspension is all the rage. You can buy Mustang II type cross-members converted to air, or you can buy systems designed for air suspension from the start. Kits should include everything you need including A-arms, bags, a reservoir, inside control, hose and the small 12 volt compressor. Art Morrison

In addition to the ability to drop your car into the weeds at will, air spring suspension provides the opportunity to fine-tune the suspension as you drive. "Basically people can play with the system as they drive," explains Art Morrison. "Once they find the exact spot where they really like the way the car feels, all they have to do is note those pressures for future reference."

Not all air suspension systems are the same. Some allow more ride height adjustment than others and different companies have different ideas about what happens when you lower the car all the way. Heidt's provides a compression stop in their system (which most standard Mustang II systems do not have) while Art Morrison explains that as you let air out of the Firestone air springs used on their suspension the air spring forms an internal cushion that acts as a suspension stop.

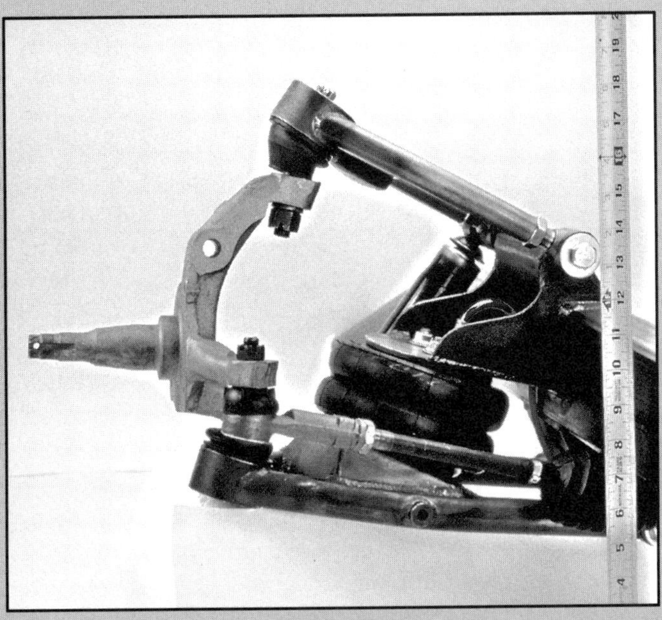

The Art Morrison front suspension uses an air bag manufactured by Firestone, designed to collapse down onto itself and provide "bump stop." Art Morrison

The Morrison system allows a full five inches of ride-height adjustment, with little or no change in camber or toe in. Art Morrison

No matter how it's done the system needs a way to allow full compression without damage to any of the suspension components. Because the basic suspension geometry isn't the same on all these systems some experience more camber change with height adjustments than others. And toe-in measurements can change with height as well.

Art Morrison's rear suspension system is based on a four-bar rear suspension with sleeve-style air spring used in place of double-convoluted style used in front. This simple effective system allows for over six inches of height adjustment.

If you're wondering how long it takes to go from fully deflated to ride height, Art reports that, "it depends on the car. For most street rods it takes about 30 seconds, but one of our employees has a '58 Caddy and that takes about two minutes to inflate just because it's such a heavy car."

Regardless of which system you buy, they all suffer a few drawbacks. First, there's all the extra hardware as compared to a standard coil spring system. Not just the air springs themselves, but the compressor, air tank, control and the lines and fittings. All of which raises the price and makes the whole thing more complex.

Second, it's hard to achieve the good looks of a polished coil-over with a black air bag. Until someone designs a way to chrome plate a rubber air spring, or hide the air springs up under the car, these systems are best suited to the fat fendered hot rods.

No two street rods are quite the same, and these air springs aren't for everyone. When you compare advantages and disadvantages, however, air springs start to look like the next big advance for street rod suspension. When you look at all the advantages air springs offer it isn't surprising that this new technology has arrived. The surprising thing is the fact that it took so long to get here.

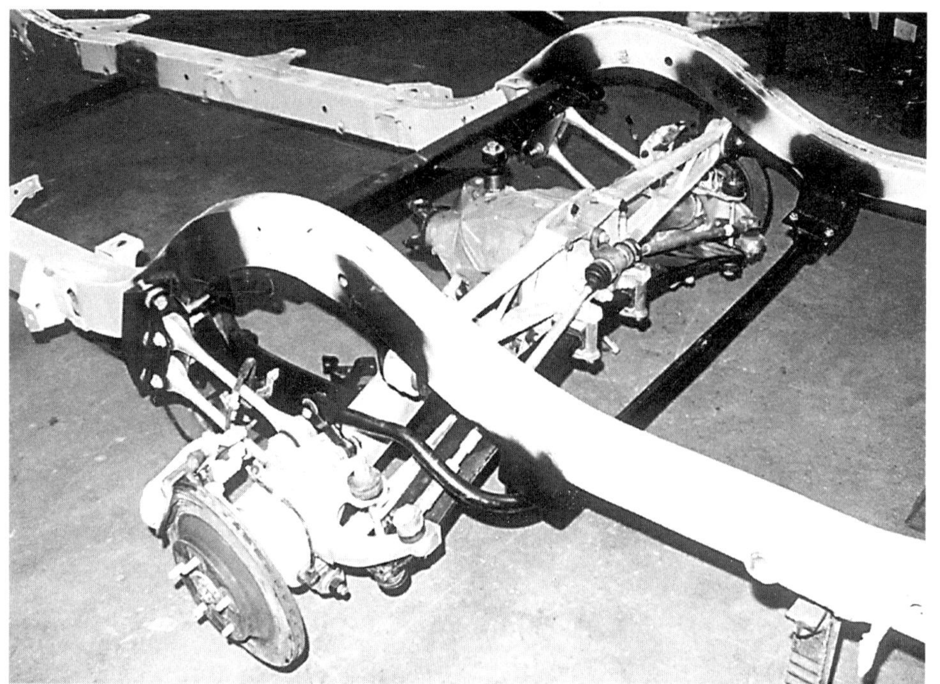

Corvette rear suspension kits are becoming more common and make it easier to adapt independent suspension to your hot rod. Because these later 'Vette systems are fairly wide, they're better suited to bigger cars like this '55-'57 Chevrolet. Fat Man

points in mind: Buy from a well-known manufacturer, longer ladder bars (and panhard rods) are better than short ones. Look for adjustability in the brackets.

Always ask the manufacturer of the suspension kit (any kit) where their parts place the tire in the fenderwell (front to rear). The factory wheelbase on some of the old machines often puts the wheel slightly ahead of the center of the fender. Just be sure the wheel and tire end up positioned exactly where you want them when the project is all finished.

Most independent systems are based on either a Corvette rear suspension or a Jaguar rear suspension. The Corvette suspension comes in two basic designs, the early design offered on 1963 through 1983 Corvettes and that used on 1984 and later 'Vettes. Note: Prior to 1980, the housing was made from cast iron. These housings are stronger than the later housings, and a greater number of rearend ratios are available as well.

Close up shows the brackets that must be added to the frame rails in order to run the 'Vette suspension. Fat Man

Kits are available from Chassis Engineering as well as other manufacturers to allow you to install your own independent suspension system, based on Corvette components, in many early and late street rods. In addition, a number of companies offer their own independent suspension system based on the venerable nine-inch Ford rearend.

Q&A: Roy Brizio

Q: Roy, they call your Dad the "Rod Father." can you give us a little background and explain the kinds of work you're doing in the shop today?

A: When I was growing up my Dad was always into hot rods. He started a shop called Andy's Instant Ts in 1966, and ran it until 1975. In 1977 I started my own shop, Roy Brizio Street Rods and we've been at it ever since. We do a lot of different types of work in the shop: we build chassis, from 28 to 40; we sell parts mail order; we have a display area up front; and we built complete cars. We also do repair work on Hot rods. We do everything.

Q: When it comes to rear suspension, how do you feel about an independent vs. a solid axle. How do you and the customer make that decision?

A: Simplicity is where street rodding is going. There are independent systems out there, we install them and they ride nice. But in my opinion most guys don't want stuff they can't fix on the road. You can get a great ride in a street rod with coil-over rear suspension and a solid axle. The coil units are so nice, there's plenty of travel, you get a nice ride and the cost is pretty reasonable. Independent systems are five thousand dollars extra and most guys have the cars so low that you can't see anything. And when the cars are really low there isn't enough travel to get the full benefit of the independent suspension. If the car is higher and you have full travel then maybe an independent makes more sense.

Q: How much harder is it to install an independent suspension correctly, so all the parts work as they should and the alignment angles turn out correct?

A: It is harder and it is more work. It's harder to set the ride height center the tire in the fender and get the pinion angle right. There are no standard measurements. When we install an independent system, we tack-weld the cross-member in place and install the parts. Then we bolt on the body and mount the fenders to ensure that everything is right and that the tire is in the center of the fender.

We install both systems all the time but the independent is still more work than a solid axle.

Q: If you make the decision

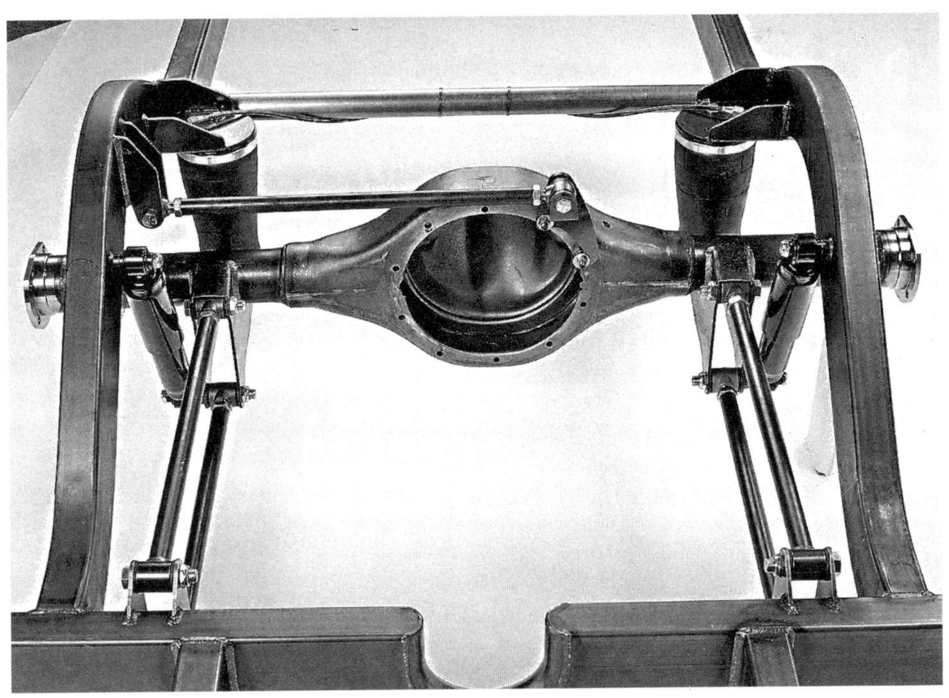

Air suspension is available for the rear of your car as well. The kit shown includes the rear sub-frame and relies on typical four-bar links combined with sleeve-style air bag that provides over six inches of adjustment. Art Morrison

Leaf-spring kits are available for most popular cars. Shown is a kit for '35 and '36 Chevy Master car that incudes Slider Springs and all necessary mounts and hardware. Available for many popular rearends. Chassis Engineering

This chassis uses the seldom-seen-on-street-rods, three bar suspension. Like a four-bar with one missing upper link, this system is known for good traction and a simple installation.

to go independent, which system do you like and why?

A: We use either a Kugel or CWI, (Concours West Industries) system. With Jaguar assemblies we find that by the time we rebuild everything, buy the new U-joints and have it all polished and plated we could just have bought a whole new set up from Kugel or CWI.

Q: For people who want to run a solid rear axle there are a number of ways to go. Some systems use leaf springs and some use coil-overs. Do you build street rods with leaf springs and do you still use the buggy spring in back?

A: I recommend leaf springs on the '35 to '40 cars because they're heavier cars and the leafs are longer so the ride is better. On earlier cars I prefer to use a coil-spring system. We do install a few buggy springs in the rear, usually on real nostalgia cars. You can make a buggy spring work but you need exactly the right spring.

Q: There are a number of different coil-spring type suspensions available. Is there one you like better than the others?

A: Everyone has a theory as to what works best but they all have their points. The triangulated four-bar doesn't need a panhard par. With ladder bars you bring all the torque to the center of the car, the torque acts on the center of mass the way Henry liked it. Ladder bars are good for big-horsepower cars especially if the car has a clutch. A straight four-bar, however, makes it easier to run the exhaust system. Ladder bars or four link, either works fine, we do about half and half.

Built at Metal Fab in Minneapolis, the three-bar uses a Panhard rod just like a straight four-bar set up.

Q. You mentioned cars with a clutch. Are four and five-speed transmissions getting more popular with hot rodders and how do you typically run the clutch linkage.

A: Yes, at least half the cars

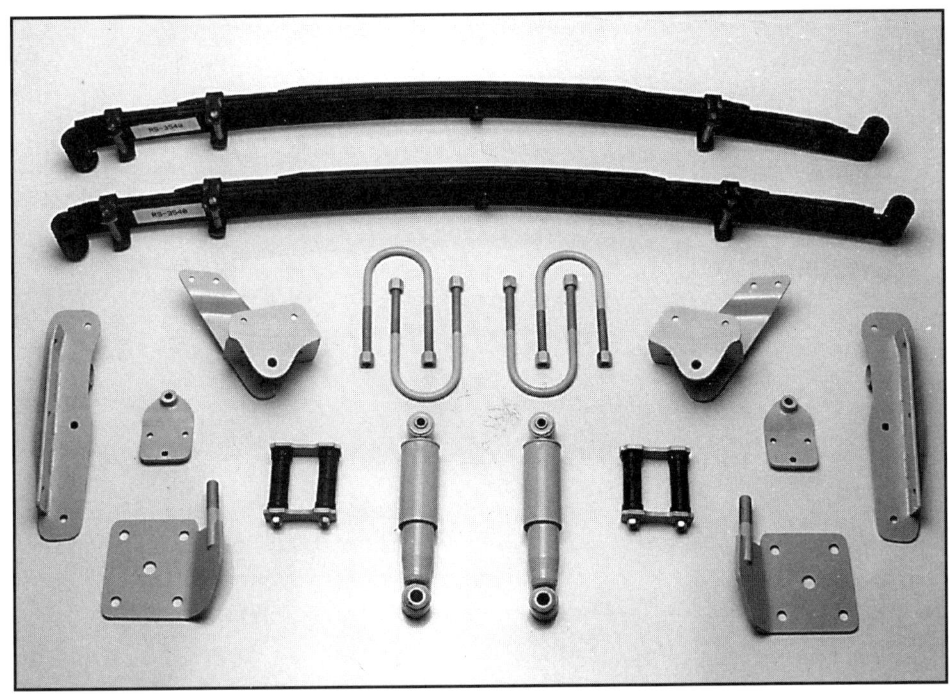

If you want to run leaf springs in the back of your rod, consider a kit like this with new springs (some come de-arched), shocks and all necessary brackets. Chassis Eng.

we build now have a clutch pedal. Part of it is nostalgia. Guys just like the stick, the last three cars I've had for myself have been stick transmissions.

When it comes to the linkage you can either use a straight mechanical linkage like the old cars had, a cable of some kind, or a hydraulic set up. A straight mechanical system is nice and simple but in some cars there isn't much room for the linkage because the column or headers are in the way.

The hydraulic systems with a hydraulic slave cylinder and throw-out assembly work OK but if something goes wrong you have to pull the transmission to fix it. You can also use a slave cylinder mounted outside the bell housing, McCloud and Wilwood both sell kits and components for this.

When we build a car with a Ford bell housing we like to use the Mustang cable system. Inside we use a hot-rod clutch pedal and pivot. We attach a heim joint to the upper end of the Mustang cable and use the stock bell-housing pivot, arm and throw-out assembly. It's a nice system and you can throw a spare cable in the trunk just in case.

Q: Which rearend do you prefer and why?

A: Well, we usually use an eight or nine-inch Ford rearend and we buy them from Currie. We just figure out how wide they need to be and then call and order it. The nine inch is stronger and there are more gear ratios available too. The eight inch isn't as strong and there aren't as many gear sets available. But, if you don't need the strength and a standard ratio is OK the eight inch is cheaper.

Q: What are the mistakes people make when they order and install a rear suspension system?

A: The only mistake is trying to save money. They try to take all the parts out of a wrecking yard and in the end they could have called Currie or one of the other vendors and ordered the parts. That way they fit the first time and it's less hassle in the long run.

Many of the best shocks use aluminum bodies and offer-adjustable valving so you can fine tune the ride to suit your tastes.

Chapter Five

Brakes

Slow and Show

Heat machines

It's hard to discuss hot rod brakes without a parallel discussion of physics. Brakes are basically heat machines. When you step on the pedal the brake pads are forced against the spinning rotors, which slows everything down. Drum brakes accomplish the same net effect by pushing the brake shoes against the spinning brake drum. In either case you are converting the car's moving or kinetic energy into heat energy.

Kinetic energy, weight and speed.

Kinetic energy depends on the vehicle's weight and

These Wilwood four-piston aluminum calipers, mated to ventilated rotors, provide plenty of stopping power for the Steve Moal roadster seen under construction farther along in the book.

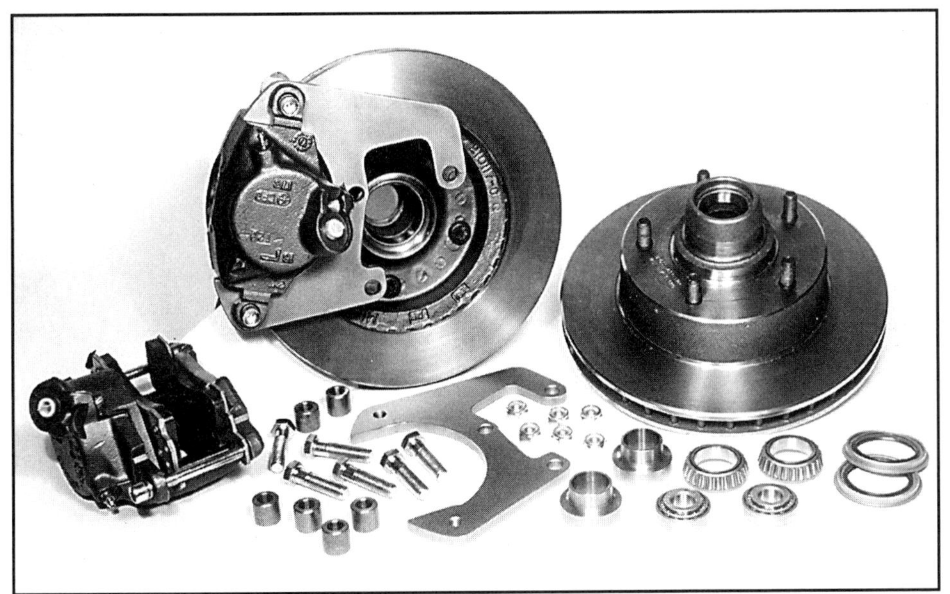

Upgrade front brakes on your '37 - '48 Ford, or '37 - '64 Ford Pickup, with big brake kits. Rotors come from late model F100s while the calipers are from '71 - '76 GM big cars. Bolt to stock Ford spindles without modification. ECI

speed. If you double the vehicle's weight, you also double the kinetic energy. If you double the speed, however, you get *four times* the kinetic energy. A hot rod traveling 60 miles per hour has four times (not two) the kinetic energy of the same machine at 30 miles per hour. So when you nail the brakes at 80 or 100mph you're asking a lot of the system. It only makes good sense to buy the best components you can for that new hot rod.

When buying brakes for a high performance automobile, more is usually better. More rotor surface area and larger calipers, with bigger pistons and brake pads.

Bigger rotors work to your advantage three ways: By providing a larger total surface area, by providing more leverage for the caliper to work on and by providing more total mass to handle more total heat. A ventilated rotor usually has more mass to handle the heat and the ventilation running through the center of the rotor helps keep it cool. A larger caliper (within reason) provides a larger piston and pad to better grab that big rotor and more mass to deal with and absorb the heat.

So far the discussion has included only disc brakes. That's because disc brakes are much better than drum brakes in most situations. Deuce coupes look good with finned Buick brake drums on the front (and those big Buick brakes do a fine job of slowing down a little coupe) yet most of us want disc brakes, especially on the front on our cars.

Disc brakes have a number of advantages over drum brakes. First, disc brakes put the friction surface right out there where the air can get at it and keep it cool. Second, any dirt or water that hits the spinning rotor tends to be thrown off, which means disc brakes are essentially self-cleaning. Third, disc brakes offer a greater surface area for a given amount of weight as compared to drum brakes.

As an interesting note the So-Cal Speed Shop manufactures a front disc brake set up that comes complete with "covers" that look for all the world like a finned aluminum brake drum.

It's hard to talk about the brakes without talking about hydraulics and the basic

The perfect blend of traditional looks with modern performance is offered by this SO-CAL front brake. Under what appears to be a finned brake drum is the Wilwood caliper squeezing an eleven inch rotor. SO-CAL

Polished hubs combined with 11 inch vented rotors make a good combination of form and function for the front of your hot rod. Heidt's

hydraulic laws at the same time. Specifically, you need to keep in mind two facts:

1) Pressure in the brake system is equal over all surfaces of the system.

2) A fluid cannot be compressed to a smaller volume.

Thus the brake fluid is not compressible, which means the pressure at the master cylinder outlet is applied fully to the pistons in the calipers. The pressure is not "used up" compressing the fluid link between the master cylinder and the calipers. This also means that the pressure at the master cylinder outlet is the same pressure that is applied to all the surfaces in the brake system.

When you buy a master cylinder you need to ensure it is matched to the other components in the system. Which brings us to a short discussion of hydraulic ratios.

A demonstration of hydraulic ratios will help explain the need to correctly match the master cylinder with the calipers or wheel cylinders. The pressure of the hydraulic fluid at the master cylinder outlet is determined by that old formula from high school: Pressure = Force/Area. If you put ten pounds of force on the master cylinder piston with one square inch of area you have created a pressure of 10psi. If you apply the same amount of force to a master cylinder with only 1/2 square inch of piston area, then you've created twice the pressure.

These street rod master cylinder come with 1 or 1-3/32 inch bore, set up for disc-disc, disc-drum or drum-drum brakes; with a seven-inch diameter booster. TCI

Assuming 10 psi of pressure in the lines and a caliper with one square inch of piston area the force on the brake pad will be 10 pounds (Force = Pressure X Area). If you double the piston area you also double the force on the brake pad. Thus the way to achieve maximum force on the brake pads is with a small master cylinder piston working calipers with large or multiple pistons having a high total area.

That's great you say, I'll run a small diameter piston in the master cylinder to create a lot of pressure and won't need a power booster. There is, as always, a trade-off in using a small diameter piston. The smaller piston doesn't move as much fluid and the pedal may be right on the floor when you've actually moved it far enough to displace enough fluid to push the pads against the rotor.

What's needed is a good match between the master cylinder and the calipers or wheel cylinders. You can't just go to smaller and smaller diameter master cylinder pistons or bigger and bigger caliper pistons. In the real world you probably want more pressure but still need to displace a certain volume of fluid.

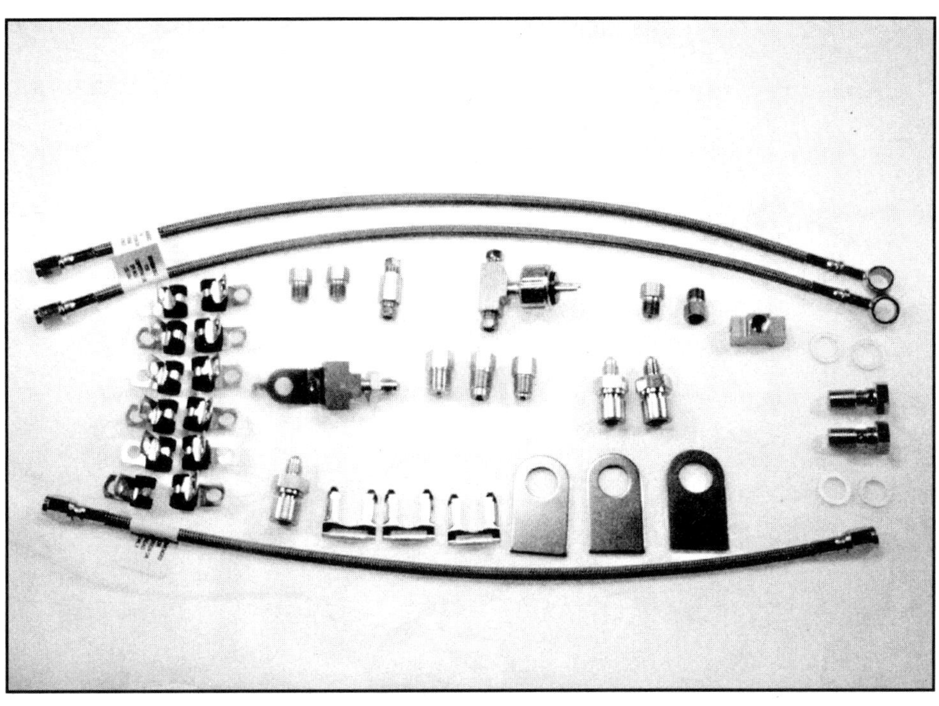

Pure Choice makes it easy to run the brake lines in your car. Just explain what it is and what you're using for brake components and they supply the parts. Pure Choice.

Detroit often uses a vacuum booster on the cars they build to provide more hydraulic pressure without any reduction in the amount of fluid displaced. For hot rods, however, you can't just add a booster to the system (more on this later) and need instead to design the system around a good manual master cylinder.

A hot rod builder also needs to consider the ideal balance between front and rear brakes (whether it's four wheel discs or not) and how that balance is achieved. Detroit uses a combination valve between the front and rear brakes to help balance out a disc/drum brake system. That combination valve does more than "balance" the brakes, in most cases it has two jobs.

Proportioning and combination valves

The basic problem with using disc brakes on the fronts and drums in the rear is two-fold. First, the drum brakes, with their big shoes, big return springs and relatively large dis-

Disc brakes with Corvette calipers are available for your street rod with most common Ford and GM rear axle assemblies. Caliper features integral emergency brake. Comes with all necessary hardware. ECI

Adapt '84-'87 Corvette rotors and finned calipers to a Mustang/Pinto front suspension. Aluminum, 6061 T6, hubs are drilled for either 4-1/2 or 4-3/4 inch bolt circle. ECI

tance between the shoes and the drum, require something like 125psi to actually force the shoes against the drums with enough force to slow down the car. Disc brakes on the other hand need only about one-tenth as much pressure to push the pads against the rotor with enough force to affect the car's speed.

So the first job of the factory combination valve is to "hold off" the front brakes until the system reaches approximately 125psi. In this way the car uses both the front and rear brakes to do the stopping, even in light applications.

The second job of the combination valve is to actually slow the pressure rise to the rear brakes on a hard brake application. Consider two stopping situations: In an easy stop from slow speed there is little weight transfer and the rear tires maintain good traction. Now, imagine the same car stopping hard from high speed. In this case there is a great deal of weight transfer onto the front tires. This leaves the rear tires with little traction and in danger of locking up. In this situation the proportioning function in the combination valve limits the rate at which the line pressure is applied to the rear brakes. The valve limits the hydraulic pressure applied to the rear brakes in order to prevent the rear tires from locking up.

What to Buy for Your Car

Before running out to the parts store or rod shop to buy the brake parts, consider first what you want, need and can afford.

No one would consciously skimp on their brakes. Most of the equipment out there, whether it's from Detroit or the aftermarket, is of good quality. The occasional problem comes from equipment that is poorly matched to the job at hand.

That job is stopping and slowing the car during street driving (remember these are *street* rods). These may be hot rods but they are not race cars. Though the race car stuff may look impressive, it doesn't always work better - or even as well - as components and systems designed for street use, and that includes good old O.E.M. stuff from Detroit.

Seen in Roy Brizio's shop, these disc brakes use a finned aluminum backing plate to improve their looks. A bright idea especially for an open wheel car.

What you buy will depend on the car's weight, bolt pattern, spindle, style and of course your budget. In general you want to buy as much brake as you can for a

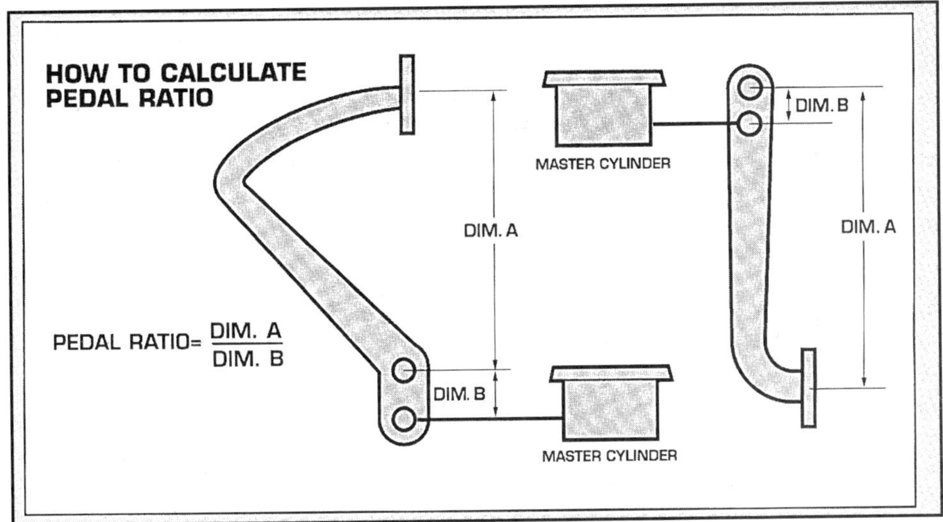

Reproduced courtesy of ECI, this diagram shows how to determine the pedal ratio, something too many of us never even consider. When you buy and install brakes on your hot rod you've got to consider all the parts - even the brake pedal linkage - and how all those parts interact.

given amount of cash. An engineer once explained to me: "When you're considering brakes, more is usually better. More surface area, more pistons and larger calipers."

As the front brakes do seventy percent (or more) of the stopping in a hard stop, it makes sense to put your best foot forward. Put the best brakes on the front. Like Detroit you may want to run discs in the front and drums at the rear.

You don't just need brakes, you must also plan and install an emergency brake. If you're using stock drum brakes in the rear then all you have to do is hook up new cables. If you want four-wheel disc brakes, the choice of rear calipers will affect your options for an emergency brake.

Rear disc brake calipers fall into two categories: Those with and those without an integral emergency brake. Most of the calipers with an integral emergency brake are from Detroit, though Wilwood now makes a rear brake caliper with an integral emergency brake.

One popular choice for Detroit-supplied disc brake calipers is the Corvette. 'Vettes have used four-wheel disc brakes for years and years, though all Corvette systems are not created equal. The calipers used on pre-1984 'Vettes have a host of problems all their own and should probably be avoided entirely. In 1984 the Corvette got a major redesign, and that redesign included new brakes. The new single-piston calipers include a saddle assembly that reinforces the caliper. These saddles must not be removed, as a way to eliminate interference between the caliper and small diameter wheels, for example.

The 1984 - 1988 Corvette rear calipers use a separate emergency brake made up of small shoes that expand against the inside of the rotor. A better choice for the street rodder looking for four-wheel disc brakes and easy emergency brakes is the 1989 and later Corvette rear calipers which feature a cable operated emergency brake built into the caliper.

Other cars that use four-wheel discs and include the emergency brake as part of the rear caliper include some Camaros, Tornados and Eldorados. In the Ford line, certain Lincoln Versailles and even

Corvette components - seen here on Jon Kosmoski's new '34 Ford -are becoming much more popular on current hot rods.

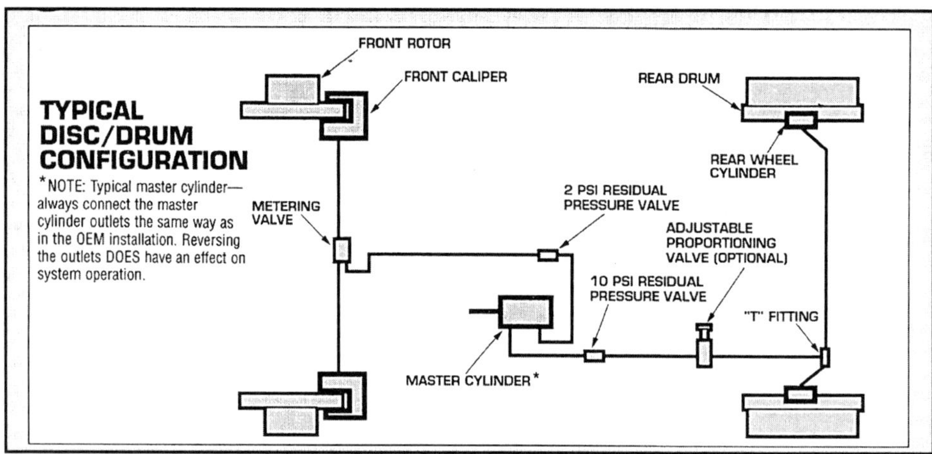

Mixing disc and drum brakes requires the right metering valve and the correct residual pressure valves. The 2 pound residual pressure valves are not always needed. (See the next illustration.) ECI

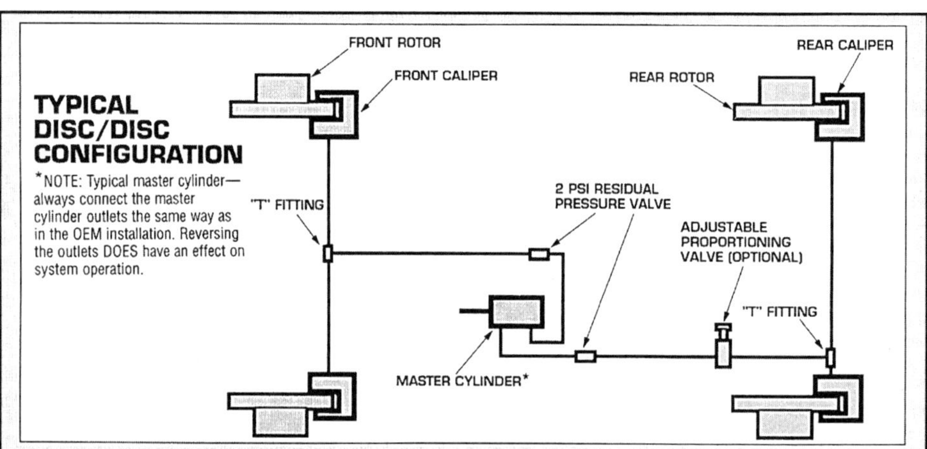

A 10 pound residual pressure valve is always used on a drum brake system. The 2 pound residual pressure valve, however, is used on a disc brake system to prevent the fluid from siphoning from the caliper to the master cylinder- only if the master cylinder is at the same height or lower than the calipers. ECI

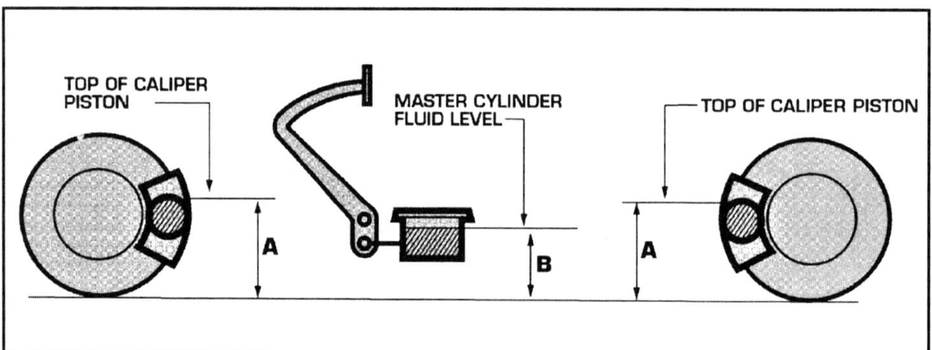

When A is greater than B the brake fluid may siphon from the caliper(s) back to the master cylinder, meaning you will have no brakes when the truck pulls out in front of your new ride. These situations require a residual pressure valve.

some Granadas used the bullet-proof nine-inch rearend with factory rear disc brakes, and these calipers include integral emergency brakes. Ralph from ECI warned, however, that the latest rear calipers from Ford, used on many Explorer and T-Birds, uses a very small diameter caliper piston and should be avoided. "That small piston means you need a tremendous amount of line pressure. Unless you're using the whole system these aren't a good choice for a street rod."

Aftermarket calipers with no integral emergency brake on the rear axle will force you to come up with your own emergency brake. Some people choose the additional rotor mounted at the rear U-joint, clamped by its own mechanical caliper. The potential down side to the single rotor and mechanical caliper is the fact that with a non-positraction rearend, if you jack up one wheel the car can roll off the jack. You can also mount additional, mechanical calipers on the rear rotors, but be sure these are good enough to handle emergency and parking duties on a 3000 pound automobile.

Use of residual pressure valves

Drum brakes need a ten to twelve pound residual-pressure check valve in the hydraulic system. This valve is usually built into the master cylinder. The ten pounds of residual pressure keeps the lips of the cups expanded out against the wheel cylinder bore and prevents air ingestion past the wheel cylinder cups when you release the brake pedal. Note: When buying a master cylinder, you not only need to buy one with the correct bore diameter, you also need to match it to the type of brakes, disc or drum, used on either end of your new car.

That same ten pound residual check valve used on a disc brake system will create brake drag. You need a check valve on a disc brake system *only* when the master cylinder is mounted lower than the calipers - the valve, a *two pound* check valve in this case, prevents the fluid in the caliper from siphoning back to the master cylinder.

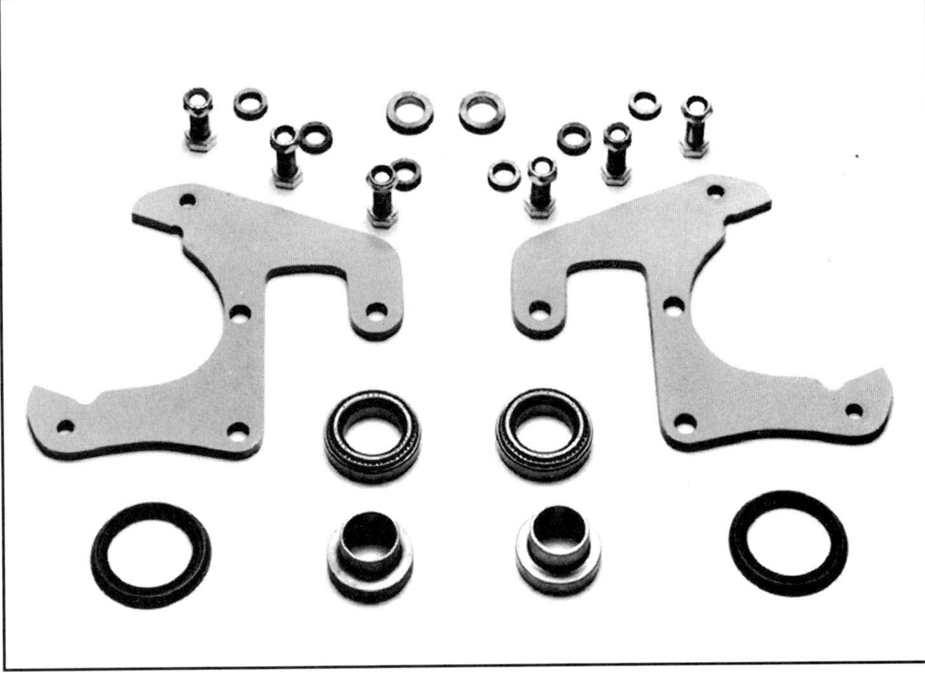

These simple kits adapt most intermediate GM calipers and rotors to a '37 - '48 Ford spindle. Chassis Engineering

Q&A: Ralph Lisena

Ralph Lisena is the owner of Engineered Components Inc. and the one man everyone calls when there's a problem with a brake installation (even other vendors of brake kits). In the Q&A that follows, Ralph shares some of what he's learned in over twenty years of designing and selling parts for hot rod and street rod brakes.

Q: What are the basic things I need to consider during the planning phase for the new car, to ensure that I get functional brakes for my new street rod?

A: The best thing to consider during planning is to make sure the brakes are big enough to do the job, Pinto brakes won't work on a 4000 pound car. What is the weight of the car you're taking the parts from. Take them off a car that's at least as heavy as your street rod.

Some of these guys think that because it's a toy it defies the laws of physics. Then they call us later wondering why the car won't stop.

Too many builders go for the pretty, showy stuff and it turns out that the showy stuff doesn't do the deed. Be sure the brake components you install came from a car of the same weight as your street rod.

You have to make sure you're using parts that make up a compatible system. If the front rotors and calipers are from a 1975 Camaro use rears from the same car or a rear system with drums and wheel cylinders of about the same size as the Camaro rear brakes.

Compatibility is of the utmost importance.

If it's a four-wheel disc-brake car the rotors on the front should be slightly bigger than those on the rear. If the rotors are the same size then the rear caliper pistons should be slightly smaller. If it's a pro/street car you might use more rear brake than you otherwise would,

These rotors and calipers adapt to nearly any Ford nine-inch housing. Polished or standard finish available for the calipers. Dutchman

Rear drum brake kits come complete with shoes, new wheel cylinders and all necessary hardware to make your life easy and minimize last-minute parts running. ECI

because of the extra rubber. The taller tire gives the wheel more leverage so it's harder to lock up that rear tire, and the width of the tire gives it more traction. But in the rain that situation can get very scary. In the case of unusual cars like a pro/street car the builder should call an expert for a recommendation.

Q: How do I make sure the master cylinder has the right bore diameter?

A: As we already discussed, you need to use compatible parts. If all the rest of the parts came off that 1975 Camaro, then use the Camaro master cylinder, or one with the same diameter as the Camaro master. You have to consider that a bigger piston displaces more fluid but doesn't provide as much pressure (for a given pedal effort). I tell people to use a bore diameter the same as what the manufacturer used for a manual non-power brake version of the car that provided most of the brake parts. Even if you intend to use a power booster, use the bore diameter for a manual brake system, because the street rod power brake boosters have a smaller diaphragm and don't provide as much boost as a factory booster.

This E-brake assembly mounts to the standard Ford pinion housing and comes complete with a cable assembly to attach to your existing parking brake cable/handle. ECI

Q: When do I need a power booster?

A: First, you should not *need* a booster. A booster is installed to make stopping easier on your leg. The only real exception is for people who use later model Corvette calipers. These use small pistons and hard pads. Especially when these are cold, they require a booster. Otherwise if the system is correctly designed you should not need that booster. It comes back to the way you design a system. Correctly designed, the manual system will work just fine though it will require more leg muscles.

People should not use a booster to solve a problem in a

poorly designed system. If you can't stop this car with reasonable effort as a manual system (with the exception of the 'Vette brakes), something is wrong. A booster is not a way to solve a problem, it should only be used to make it easier on your leg.

Q: Should I use factory or aftermarket calipers?

A: Either one will work fine in a correctly installed system, though the true race car stuff can be a headache because it wasn't designed for street use. But otherwise, the Wilwood or other aftermarket calipers will work fine if the system is correctly designed. But you need to consider serviceability. Can the caliper be repaired or replaced if it breaks down in the middle of Montana? You can get GM parts anywhere, even on a Sunday afternoon. Remember, no matter what they tell you, shit breaks.

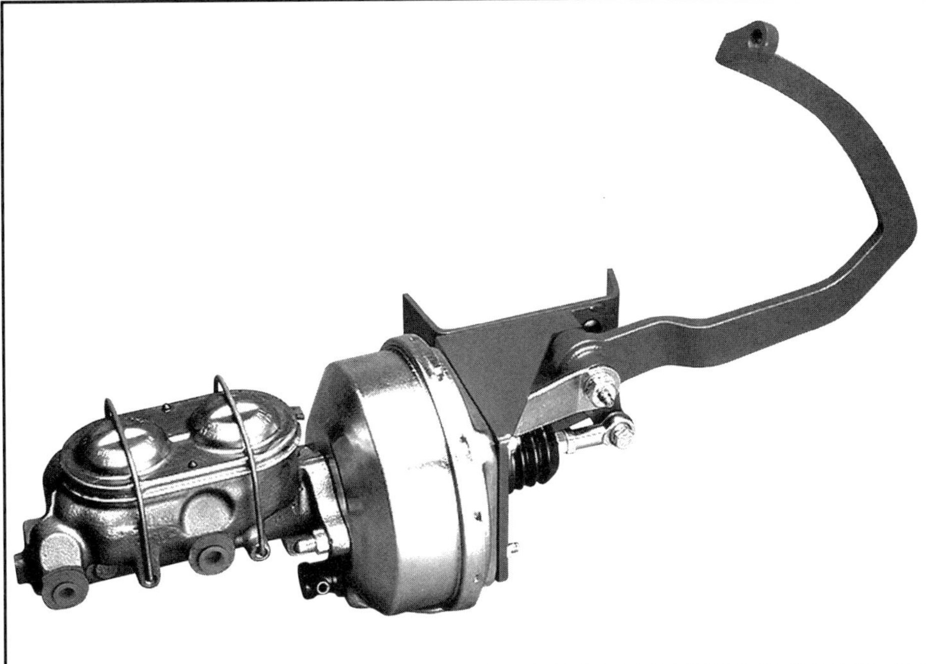

This dual-chamber master cylinder from TCI comes as an assembly with the seven-inch booster, the brake pedal and the correct pedal-mounting bracket. TCI

When you use the aftermarket calipers consult the manufacturer. Ask them for a recommendation regarding the pedal ratio and master cylinder bore size. It may be hard to get an engineer on the phone. If the caliper has two or four pistons that has to be figured into your overall design for the brakes.

The non-floating calipers must be perfectly aligned to the rotor. This is harder than you might think. Factory floating calipers are more forgiving. The caliper can square itself up when you apply the brakes. With rigid-mount calipers the pads wear at an angle and the car requires tremendous pedal pressure to stop.

Q: With a disc-drum system, can you talk about the need for a hold off valve or factory metering valve?

A: First, with any car you always want to apply the rear brakes first, it provides stability especially on slippery surfaces. Factory four-wheel disc brake cars usually take care of this through the master cylinder design.

With a disc-drum system you must have a metering function because drum brakes take 100 to 125psi to overcome the return springs and move the shoes up against the drums. Discs, however, only need 10psi to move the pads up against the rotor. On a disc-drum system without some kind of metering valve, you use only the front brakes at lower speeds.

The combination valve used on Detroit's disc-drum

This front disc brake hold-off/metering valve is used for the front brakes in a disc/drum braking system to provide a "hold off" feature so as to allow the rear drum brakes to actuate first. This function is essential for correct system operation. ECI

This remote brake booster is intended for cars too low for even a 7 inch brake booster. Mounted in the trunk this unit works with any dual remote master cylinder. Not to be confused with remote boosters designed to work with single chamber master cylinders. An electric vacuum pump is available for cars with radical engines/low vacuum. ECI

cars actually does two things. It *holds off* the front brakes until the pressure is high enough to apply the rears, and it reduces the rate at which the pressure is applied to the rear wheel cylinders during a stop. This second function is done to prevent the rear wheels from locking up when much of the weight has shifted to the front wheels.

But it's hard to recommend one of these factory valves for a street rod because we don't know the weight of the street rod, the diameter and width of the tires or the design of the brake system and how all those numbers compare to the Detroit car the valve assembly was designed for.

We (ECI) now build a metering valve with the *hold off* function only. There is no provision to limit pressure to the rear brakes. We've done this because our testing shows that on the typical street rod the tires are so tall and so fat they don't lock up anyway. We recommend that the customer use an adjustable valve in the line to the rear brakes if the car is really light in back.

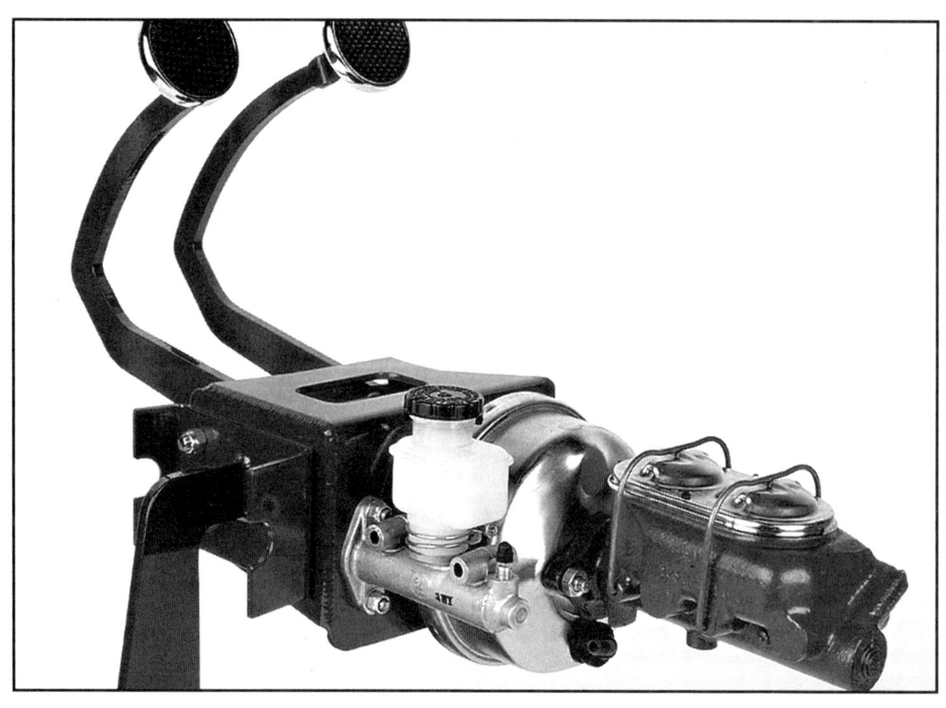

It doesn't have to be an automatic transmission. Assemblies ike this are available that provide a dual chamber master cylinder for the brakes and a clutch master cylinder for hydraulic clutch actuation. TCI

Q: What are the mistakes that people commonly make when they design and install brakes on their street rod?

A: They use single-chamber master cylinders (instead of split or dual-chamber master cylinders). They mis-match the front and rear brakes, as we discussed earlier, maybe because they just don't understand that all the parts of the brake system need to work together. I tell people that it's a good idea to talk to people who have the same components you intend to use. Find out who they bought it from, how does it work, and are there any problems. And never listen to your buddy, he knows less about it than you do on any given day.

These guys forget to make sure they can bottom out the

master cylinder when there is no fluid in it - before the pedal hits the floor. If you can't bottom out the cylinder you will never engage the second half of the system when that hose blows or half the system runs dry.

They put massive rear brakes in the rear of light cars. If they aren't sure what to install they should find a brake engineer (not a salesman) who can help them determine what's right for their car.

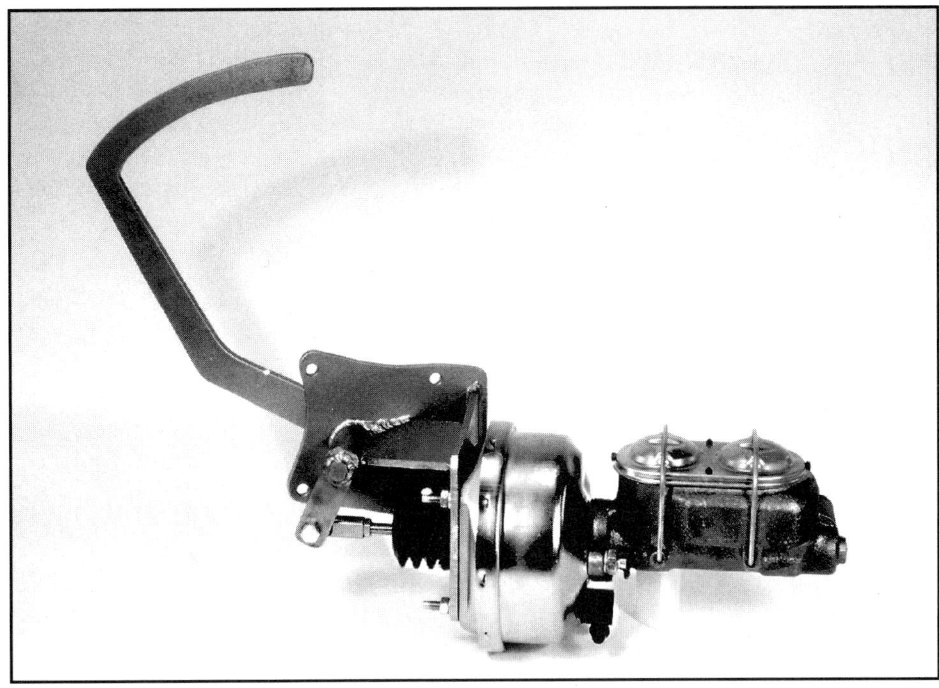

Brake booster/master cylinder assemblies are available with matching brake pedal linkage and all necessary brackets. Remember, any master cylinder you use must be a dual-chamber master. ECI

BRAKE FLUID CHOICES

When all the components are installed and you're ready to bleed the system you might figure the tough decisions are over. Then you get to the auto parts store and see at least three grades of brake fluid on the shelf. What to do?

Brake fluid is simply a very specialized hydraulic fluid. One that operates in dirty environment and must withstand very high temperatures without boiling. Boiling brake fluid is a gas and thus a compressible material. Gas in the system is sensed by the driver as a spongy brake pedal. Our fluid must stay viscous at nearly any temperature and resist boiling up to at least 400 degrees Fahrenheit.

There are three grades of brake fluid commonly available, DOT 3, DOT 4 and DOT 5. DOT 3 and 4 are glycol-based fluids with dry boiling points of 401 and 446 degrees Fahrenheit respectively. Either fluid is suitable for use in disc brake systems. There are two basic problems with DOT 3 and DOT 4 brake fluids: They tend to absorb water from the environment (they are hygroscopic) and they attack most painted surfaces.

Glycol-based brake fluid containers must be kept closed so the fluid won't pick up moisture from the air. Because the DOT 3 or 4 brake fluid in your car will pick up some water no matter how careful you are, it's a good idea to flush the system with fresh fluid every couple of years. Remember that brake fluid contaminated with water boils at a much lower temperature and can be corrosive to components.

DOT 5 fluid is silicone based, meaning a higher boiling point (500 degrees Fahrenheit, dry), no tendency to absorb water and no reaction when spilled on a painted surface (though silicone fluid can stain the paint if not washed off quickly).

This Econo-Brake-Kit adapts '77 down GM hub and rotor or Mopar hub and rotors to '37 - '48 Ford spindles. TCI

Wilwood manufactures proportioning valves for use in the line to the rear brakes, and residual pressure valves described earlier. Chassis Engineering

Like every other advance silicone brake fluid has its trade-offs. In fact, silicone isn't everyone's favorite brake fluid. Silicone costs more, it is slightly compressible, it aerates more easily than glycol-based fluid and it is said to cause swelling of the brake cups and seals after long term exposure. The choice is yours, but no matter which fluid you decide to use stick with it and do not mix one brake fluid type with another.

Tips on brake installations

To ensure safety it's necessary to follow certain procedures whenever you do any brake work. These include good attention to detail, extreme cleanliness when dealing with the hydraulic part of the system, careful examination when the work is finished to check for any leaks, and a careful road test. Solvents will attack the rubber used as seals in brake systems, so all cleaning of hydraulic parts must be done with clean brake fluid.

Inactivity is very hard on brake parts. Overhaul or replace wheel cylinders and calipers that come along with the rearend or front clip you drag out of the junk yard. Discard any old factory rubber hoses and replace them with new components. Master cylinders too should be overhauled or simply replaced with new components. Be sure any master cylinder you use is a two-chamber design.

If you force the caliper piston from the bore with compressed air be careful that the piston doesn't become an air-powered projectile. Stuff the caliper cavity full of rags and be sure to keep your fingers out of the way when applying air to the caliper. Once apart, pitted caliper pistons need to be replaced, The caliper bore should be thoroughly cleaned with the correct brake hone and the groove for the main piston seal must be cleaned thoroughly. The new seal and the piston should be lubricated with brake fluid, or brake-assembly-fluid, before being installed.

If you've never overhauled a brake caliper before it might be easier and simpler to buy rebuilt calipers from your local auto parts store or one of the street rod suppliers.

As mentioned, most O.E.M. calipers are single-piston designs that float. If the pins or the sliding surfaces are dirty

Many Ford and GM rear axle assemblies can be converted to disc brakes with this kit. Included are GM rotors mated to '80-'85 Caddy Seville calipers with integral E-brake mechanisms. ECI

and rusty the caliper can't float. This means you must be sure to clean all sliding surfaces, and replace the pins on G.M. calipers if they're rusty.

Always have used rotors or drums turned on a lathe and checked against factory specifications to ensure that there is still enough material left after turning. Be sure to clean all the old grease and metal filings out of the hub, inspect the bearings and races, pack the wheel bearings and use a new seal.

If you're rebuilding the old drum brakes, the hardware and springs that hold and retract the shoes should probably be replaced at the same time you're doing all the other work. Many good automotive parts stores and some street rod vendors sell brake hardware kits for most drum-brake applications.

Caliper brackets should be original or come from a good aftermarket supplier. The full force of a panic stop is transmitted to the chassis through that caliper bracket so don't skimp. Use a good bracket and bolt it to the spindle assembly with the hardware supplied with the kit or with grade-eight bolts.

Many of the popular front brake kits for the Mustang II and even some early Ford axles combine an eleven inch ventilated rotor with the intermediate or large GM caliper. The rotor used in some of these kits is thinner than the stock G.M. rotor (most of these rotors are .810inch thick while the original was .960inch) and that's why some of these kits also supply a spacer to be used behind the inner brake pad. If you leave out the spacer the piston comes out farther than the GM engineers intended. This might be all right until the pads become worn - and the end of the piston is pushed past the inner seal and you lose the front brakes.

Non-floating aftermarket calipers come with a series of small spacers. You will need to use these to center the caliper as it mounts over the rotor. That way each piston moves the same distance on a brake application.

Mount your new or rebuilt caliper to the caliper bracket and get ready to hook up the hoses. Hoses should be new and carefully chosen to ensure they are the correct length. It's easy to install a hose that's too short or too long - a hose that will tear on a bump or rub on a tire. After installation run the suspension up and

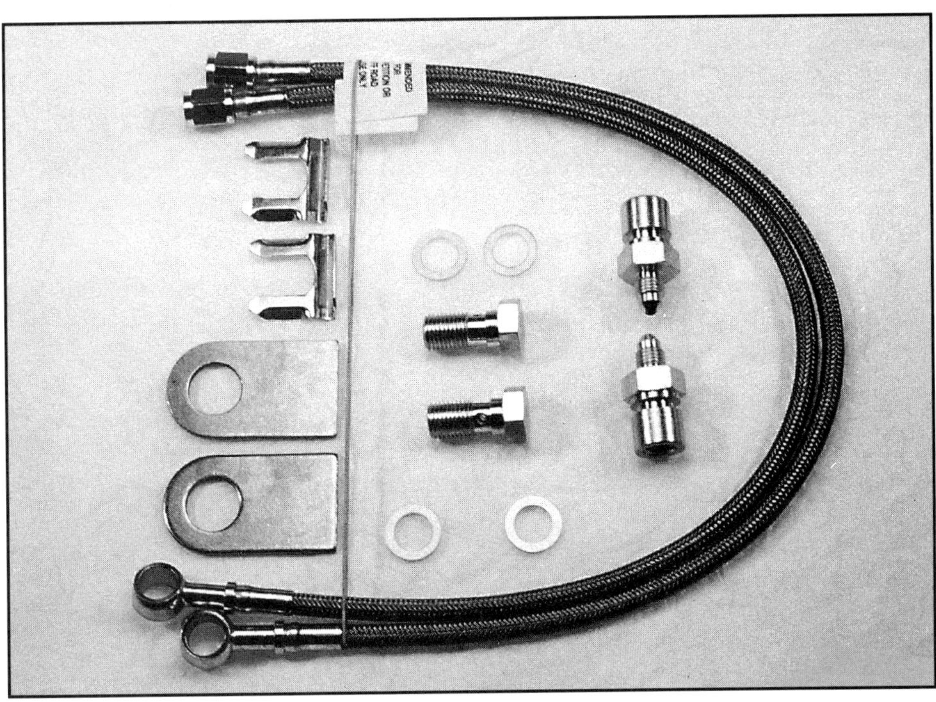

Some calipers use metric fittings, which require the correct banjo bolt and/or brake line. Any line you use must be the correct length and be well supported by brackets so as not to rub against the tire or suspension components. Pure Choice

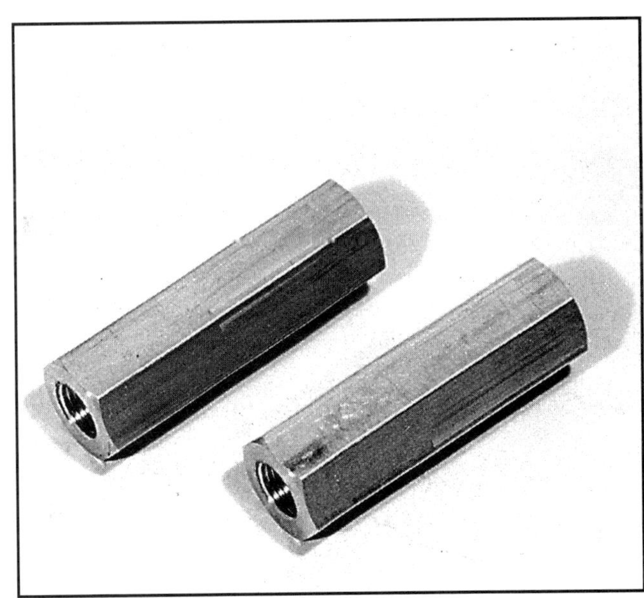

As explained earlier, the 2 pound residual presure valves are used as an anti-sipphoning valve on some disc brake applications; while the 10 pound valves are always used in drum brake applications if the master cylinder does not have a built in residual pressure valve. ECI

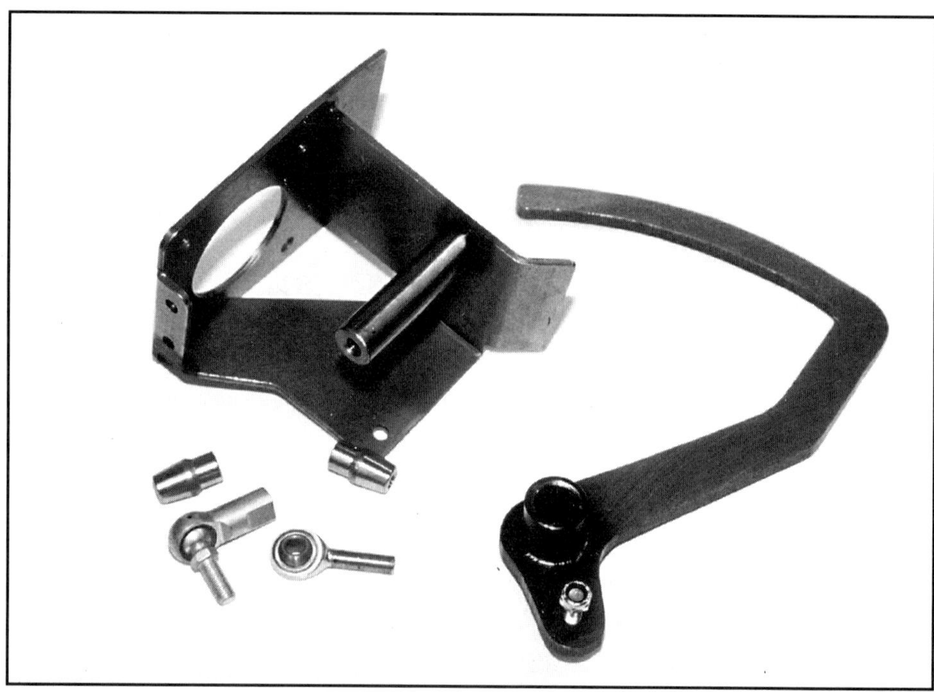

Raw pedal assemblies are available from a variety of sources, with brackets designed to fit most frames and X-members.

down, and back and forth, to check for any potential clearance problems.

When it comes to flexible hoses, braided lines look great and may be stronger than stock flexible hoses. Most, however, are not DOT approved and could cause your car to be failed during a state inspection.

Be sure that the bleeder screw on your new calipers ends up at the top of the caliper. If not, you will have to bleed the brakes with the calipers off the bracket. When you're done bleeding have someone put plenty of pressure on the brake pedal and then check all the connections and bleeder valves for any leakage.

Some builders run the brake lines inside the frame rails. While that probably constitutes overkill, it *is* important to use clamps to anchor the brake lines to the frame rails. Loose brake lines vibrate, crack and rub on the chassis. Grommets must be used where the line passes through an opening in the frame or body and lines must be kept away from any source of heat.

A panic stop can generate as much as 1000psi in the hydraulic system - too much pressure for anything but an approved steel brake line, with double-flare fittings or systems specifically designed for automotive brakes. Stainless steel lines are available, though they require the use of 37 degree AN fittings in place of the standard 45 degree double flare fittings.

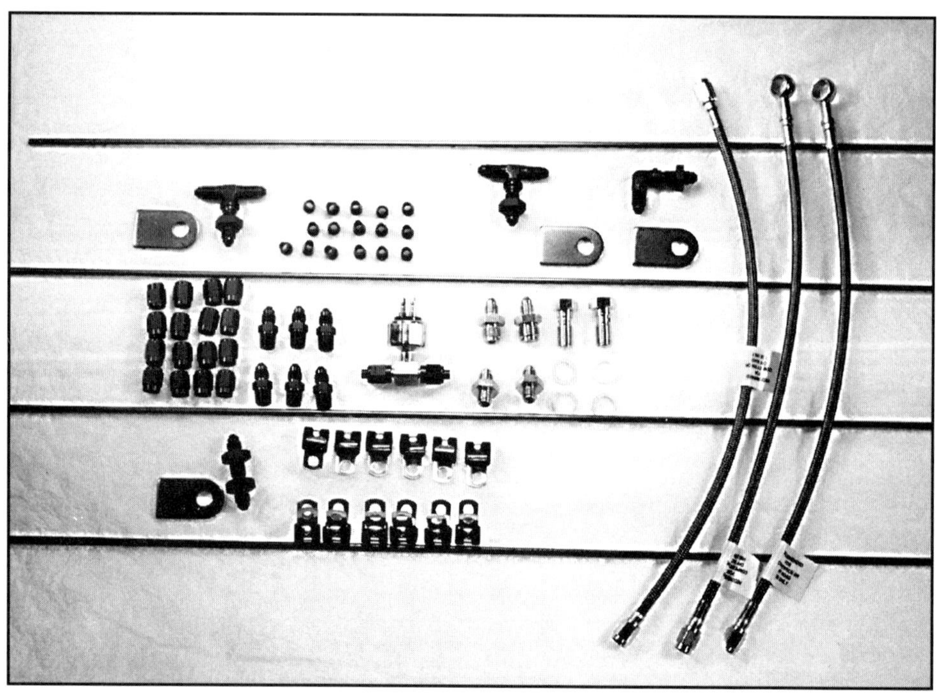

Pure Choice makes available complete brake line kits for nearly any imaginable application. Order either the standard 45 degree fittings or the 37 degree fittings used with stainless brake lines. Pure Choice

Mounting the master and brake pedal

Most street rods mount the master cylinder to the frame. This keeps everything mounted low, and all that hardware off the firewall. A variety of mounting brackets are available, or the enterprising builder can fabricate his or her own.

The master cylinder bracket needs to be sturdy, so the full

movement of the pedal is transmitted into piston movement and not in flexing of the mounting bracket. Many builders mount the bracket solid to the left frame rail and then find a way (easier on some cars than others) to tie the bracket to the X-member or one of the crossmembers. While the master cylinder and booster need to be below floor level, you don't want them any lower than necessary.

The dimensions of the pedal assembly determine the pedal ratio, another of those pesky details to be considered when planning out the brake system (see the illustration for a better understanding of the dimensions and how they affect the ratio). A high pedal ratio will provide tremendous leverage for your foot, allowing you to generate high line pressure, though the pedal-travel necessary for a given application will increase. Conversely, a low pedal ratio will decrease the leverage; meaning an increase in the leg-force needed to generate a given amount of pressure and a decrease in the pedal-travel. Ralph from ECI recommends a ratio of about 4.75 or 5 to 1 as the best for master cylinders with a bore of 7/8 or 1 inch.

As mentioned earlier, it's a good idea to mount the dry master cylinder on the bracket to be sure the pedal can move all the way to the end of its travel without hitting the floor or some other obstruction. Before you do the final mounting of the master cylinder always "bench bleed" the master cylinder. Just use your fingers as one-way valves, allowing air and fluid to push past your finger tips when the pushrod is moved into the cylinder and sealing the outlets as the pushrod is allowed to come back to its rest position. When you mount the master in the car the job of bleeding the brakes will go much faster because the master cylinder has already been bled.

If you use a vacuum-operated power booster be sure to use a one-way valve in the vacuum line. Not only will this provide a more constant supply of vacuum to the diaphragm, it will keep gas vapors from seeping down the line (remember, they're heavier than air) and turning the brake booster into a potential bomb.

Be sure the centerline for the pedal pivot is perpendicular to the centerline of the car so the pedal moves straight and not through an arc. Remember that the master usually ends up mounted backwards from the way it was mounted in a Detroit car, so the hoses and reservoirs are backwards. Be absolutely sure it's plumbed correctly, with the front half connected to the front brakes. The people who say it doesn't make any difference which way the master cylinder is connected don't know what they're talking about.

The success of your brake system comes down to compatibility and attention to detail. You must take the time to plan and buy components that are matched to the other components. Then you need to install those new and rebuilt parts in such as way that there are no leaks, no binding of the master cylinder linkage and no chance for the lines or hoses to vibrate or chafe against a sharp edge.

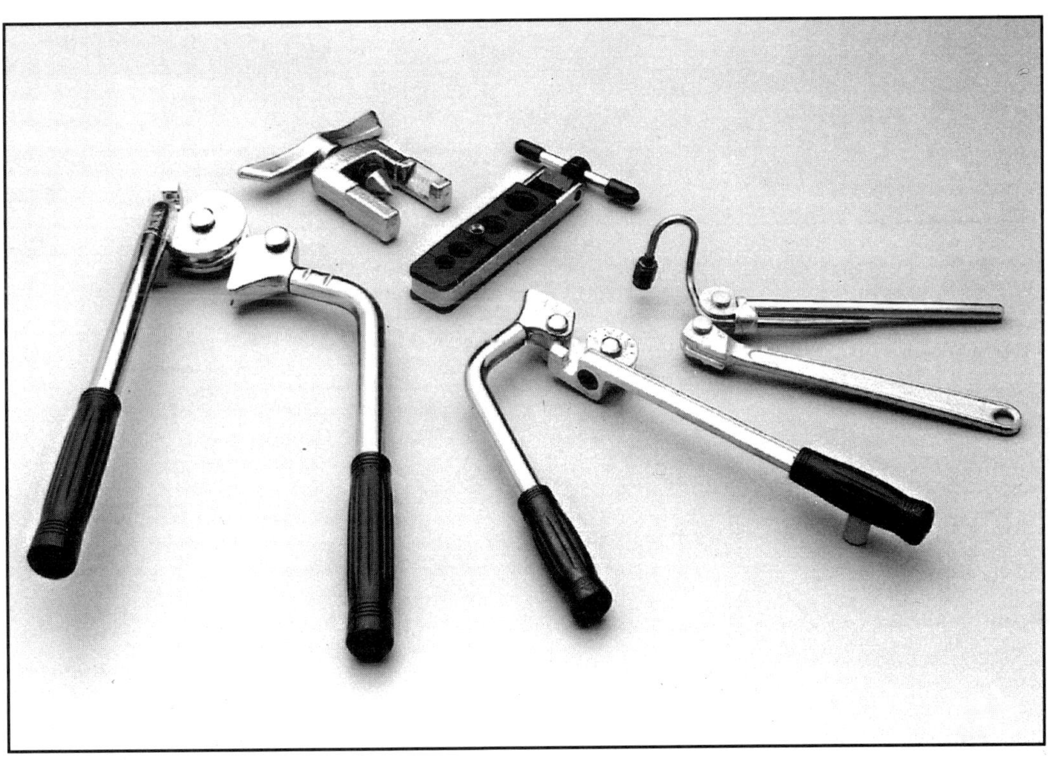

A neat brake line installation requires neat, accurate bends. Far better than bending them on the bench over an old piece of pipe, these tools make it possible to bend the brake lines exactly where and as much as you want. Pure Choice.

Chapter Six

Engines

The Heart of Your Hot Rod

PUT THE HOT BACK IN HOT ROD

Though the majority of street rods carry a small-block Chevrolet engine, there are a large number of equally useful power plants out there. If money is tight and there are no other over-riding considerations the small-block Chevy engine is the logical choice. The Chevy engine is durable and plentiful, all of which makes the little bow-tie the best bang for the buck.

Certain cars, however, call out for more. A nostalgia Model A roadster like the one seen farther along in this book is tailor made for an early Oldsmobile Rocket V-8. Flatheads have a certain charisma all their own and there

A look under the hood of a very special hot rod belonging to Steve Moal, builder of the Tim Allen car seen in Chapter 10. Note the "engine turned" fire-wall, the hand-formed air cleaner and neat valve covers.

is nothing like an old Chrysler Hemi to brighten the engine compartment of nearly any ride. In fact, it seems of late that more and more hot rod builders, both professional and amateur, are choosing alternate sources of power. Nailhead Buicks, rare SOHC Ford V-8s, and even Caddy motors are turning up under the hood of many "new" hot rods.

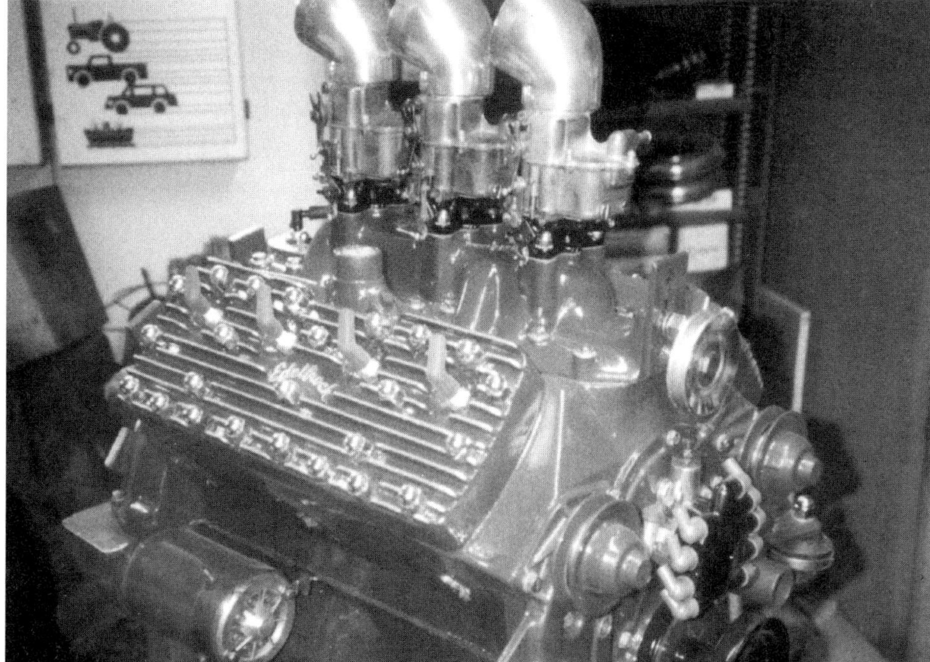

Certain engines have a charisma all their own. In spite of the extra maintenance that a flathead requires, many choose Henry's first V-8 to power their hot rod.

Essentially, you need to achieve a balance between your budget, the design of the car, your own mechanical skills and how you intend to use the vehicle. If you intend to drive the car extensively and want good power with minimal maintenance, then a GM small-block (possibly a new "crate" engine) is probably the way to go. If, on the other hand, you want to drag race the car on weekends, or you love Hemis or flatheads, the stock small-block just won't do the job.

How much is a new motor?

Most of us are on a budget and can only spend so much money on the motor for that new street rod we're building. Thrifty shoppers, we often buy used stuff because it costs less than new. The logic of buying used is hard to refute, until you consider some of the relative bargains offered by your local Bow-Tie dealer on new "crate" engines.

The only thing more trick than three deuces is six! Note the polished and painted Corvette valve covers. Kurt Senescall

Crate engine is kind of a loose term. Here it means new engines offered for sale by the various factories and usually sold through the dealerships. Though the best known crate engines come from Chevrolet, (see the Side-Bar in this chapter) Ford, and even Chrysler, offer complete and nearly complete engines.

Ford engines

Ford Motorsport SVO makes available everything from mild 302 short blocks to complete 460 and DOHC 4.6L

Not your standard small-block, installation of the seldom seen SOHC Ford engine required the bulges in the hood sides. Kelsy Petrykowski

Cobra engines. You can even buy complete 429 Boss engines with aluminum heads, originally designed to run the NASCAR tracks in 1969. Among the more popular of the Ford offerings is the 302 long block with high-flow GT-40 heads, and the 351 Windsor block with aluminum heads and a Victor Jr. intake manifold.

The SVO catalog also lists a short water pump for 289, 302 and 351W engines and another designed for serpentine belt use (reverse rotation). Also of note is the oil pan kit that moves the sump to the rear of the engine, thus eliminating interference problems with many front cross-members.

Mopar

The Mopar guys aren't far behind when it comes to this crate engine business. In fact, some people would say they're out front, in terms of cubic inches anyway. How about a 528 cubic inch street Hemi, ready to drop into the rod of your choice and terrorize the local big-block Chevy guys. If that's a little strong there are a variety of 360 cubic inch small-blocks available with ratings of 300 to 380 horsepower. And for the dare-to-be-different crown Chrysler sells a variety of complete Viper engines.

Finding the right used engine for your project

Whether the engine you intend to run will be another of the proverbial small-block Chevys, or something unusual like a Ford flathead, you still have to go out and find that certain motor. The one that is rebuildable, already rebuilt by a reputable shop, or brand new and in the crate.

Before ruling out a Hemi for power, consider that complete and assembled new Hemis are available from Mopar Performance Parts. Kelsy Petrykowski

If you're shopping for a used motor in good (or at least rebuild-able) condition, there are a number of places to look. The alternatives include the junk yard, the swap meet or the "donor car."

A junk yard engine should come with some kind of warranty or a guarantee that the block will clean up with only a .030inch overbore.

Don't buy a used engine without some assurance that the engine you buy can be rebuilt with a minimum of trouble and expense.

We all love to cruise the swap meets on Sunday morning, but it might not be a great place to buy an engine. Is the guy who sells you that nice clean 350 on Sunday morning going to be around later? What if it turns out the really good used engine lays down a cloud of smoke that makes your car look more like a mosquito fog machine than a hot rod? Rather than buy a used engine - one you probably haven't heard run, with an unknown history - why not buy a used car instead. A beat up Camaro, or Fairlane, or Impala can often be had for less than the price of a used engine.

High tech smoothies from Boyd require an engine compartment just as smooth as the body. Modern fuel injection engines lend themselves to this ultra sanitary look. Kurt Senescall

The popularity of the donor car has diminished in recent years as more and more companies offer mail order shopping on everything from engines to transmissions to rearend assemblies. But if you're looking for a good, inexpensive engine, it might make more sense to buy the whole car simply as a means of guaranteeing that the engine is indeed serviceable before you take the time to stuff it between the frame rails.

Whether you bought a whole car or just a used engine, the time comes when you need to decide exactly what to do with the new/old engine. Making an intelligent decision for the engine requires that you have a plan for the car: How fast does it really need to be? How, and how much, will it be driven?

Not a smoothie, this old ride uses an antique V-8 with lots of chrome, carburetors and sex appeal, all hanging out in the breeze for everyone to see. Kurt Senescall

Will the new fat-fendered Ford be part of next year's Power Tour, or just a weekend cruiser, seldom getting more than fifty miles from home? Do you need an engine that goes fast, or just one that looks fast?

Planning for the engine requires that you be honest, not just about the car's future use, but your own mechanical skills as well. The cost of a rebuild can be reduced considerably if you're willing to do most of the disassembly and assembly yourself. High horsepower engines, or old and unusual power plants, often require more maintenance than bullet-proof new crate engines.

You need to consider not only how much you're going to

GM Crate Engines

Bow-Tie Smorgasbord

For a little help in determining the full range of available GM crate motors I spent a few hours at Friendly Chevrolet in Fridley, Minnesota. A certified GM Performance dealer, Friendly moves a large number of crate motors and high performance parts. Their high-volume sales are helped in part by two running, demonstration engines: a 502/502 big-block and a ZZ4 350 cubic inch small-block. As Jerry Martini, Wholesale Sales Manager explained, "People come in who think maybe they want to buy a crate engine. And it helps if we can show them the compete engine, ready to run, on a stand. And if they aren't sure whether to buy a small-block or a big-block, we fire up both engines. But after hearing both engines run they almost always buy that 502."

As Jerry explained, GM crate motors start with a 250 horsepower 350 cubic inch small-block that comes in long-block form, complete with all the "tin" for around thirteen hundred dollars. The next step up is a 300 horse 350 cube crate. The final stop in this ladder of ascending horsepower is the ZZ4. By combining aluminum heads with a four-bolt block equipped with 10:1 cast pistons and a roller camshaft, this fourth-generation double-zee attains 355 horses and over 400 foot pounds of torque. As we go to press there is one more step on this horsepower highway: a limited edition ZZ 430 capable of milking 430 horsepower from only 350 cubic inches, largely through the use of some new magic aluminum heads developed by the engineers at GM.

If you're looking for a high-tech alternative to the standard small-block, then just turn the pages of the large and impressive Performance Parts catalog until you come to the LT4. The catalog describes this 335 horsepower, 350 cubic inch V-8 as a "box full of technology." What really sets this engine apart from its more mundane siblings is the cam-driven water pump. The aluminum heads are equally modern, and come with 2.00 inch intake valves and sodium-filled 1.55 inch exhaust valves. Combustion chambers feature fast-burn 54.4 cc combustion chambers that provide a 10.8:1 compression ratio. And if 335 horses aren't enough to fill your barn, the relatively mild roller camshaft, normally installed in this engine, can be swapped for a hot LT4 camshaft good for another 20 horsepower.

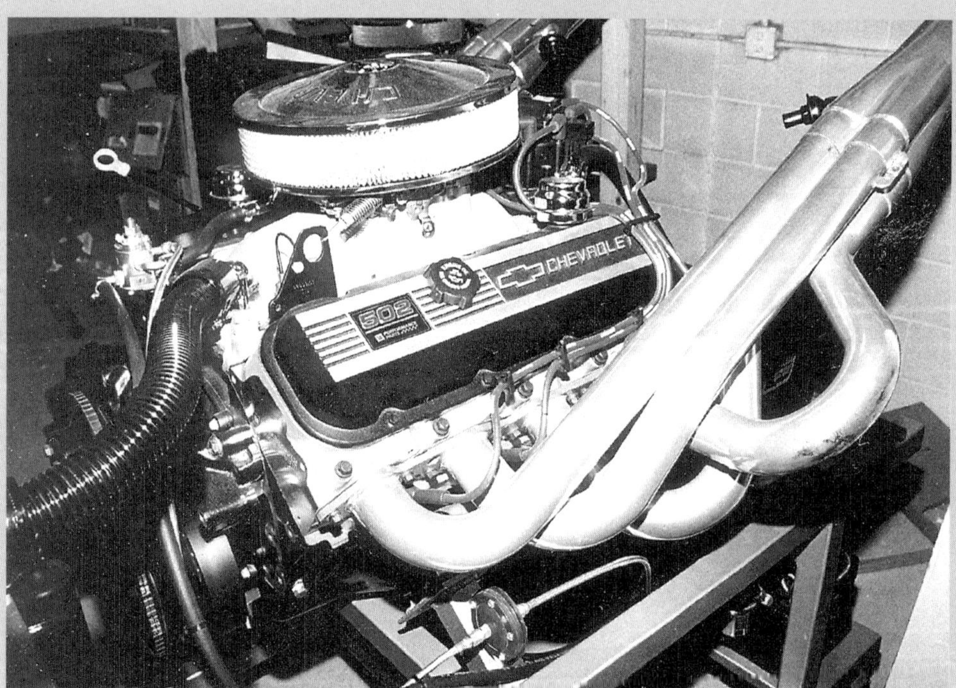

Friendly Chevrolet in Fridley, Minnesota installed two of GM's most popular crate engines on stands ready to run. The zoomey heads are bolted to the king of the big-blocks, a 502-502. These engines come assembled or as a complete kit.

For those who believe that old adage about cubic inches, the bow-tie catalog offers 454 and 502 cubic inch big-blocks in ratings from 425 to 502 horsepower. The 425 horse 454 uses cast-iron heads with rectangular ports and big valves (2.19 and 1.88 inch for intake and exhaust respectively) mated to a cast-iron block with a roller camshaft, forged crank and

GM Crate Engines

cast 8.75:1 pistons. The combination of large cubes and modest compression means this engine can live easily on the street drinking nothing more expensive than pump-premium while generating 500 foot pounds of torque at 3500 RPM.

Middle ground in the big-block side of the catalog is a 450 horsepower 502 cube HO engine with cast-iron, rectangular-port heads mated to a high performance four-bolt block. The compression is the same 8.75:1 seen in the smaller big-block, but the additional cubes mean this thumper unleashes 550 foot pounds of torque when equipped with a 750 CFM carburetor, HEI ignition, and 2X30 inch headers.

But the demonstration engine that generates all the big-block sales at Friendly Chevrolet isn't some little puny 450 horse V-8. No, the baddest of the bad is the 502/502, available as a complete long block or a complete kit. The Godzilla of street motors comes with big-valve aluminum heads (2.25 inch intakes), roller camshaft, complete induction system, ignition system, and even a high-performance gear-reduction starter. The four-bolt block is filled with forged pistons, heavy duty connecting rods and a forged steel crankshaft.

If you're a skeptic, especially where big dyno numbers are concerned, the folks at Friendly Chevrolet have already taken steps to relieve any doubts about the true output of these engines. "We ran a 502/502 on the dyno recently," explains Jerry. "And it did 510 horses on the first pull. After a little massaging they got 520 horses with the factory 850CFM carburetor. Later, with a bigger, Dominator carb, they got it all the way to 550 horsepower."

If a bow-tie crate motor sounds like the perfect no-hassle power plant for your new hot rod, cruise on down to your local GM dealer and ask to see one of the Performance Parts catalogs. Prices do vary from one dealer to another so it may pay to shop around. Dealers who are part of the Performance Parts Program are more likely to have a variety of the available engines in stock (only about 800 of the Chevrolet dealers in the country qualify as members of the Performance Parts Program). At Friendly Chevrolet they find that it pays to keep a large stock of the most popular crate engines. "We find that we sell more engines because we've got them here," says Jerry. "These are buyers who want to put their hands on the new engine, they don't want to order it from the catalog and then have to wait for delivery."

The other ready-to-run crate engine at Friendly Chevrolet is this ZZ4 small-block rated at 355 horses and 400 foot pounds of torque. You can also mix and match a short block with aluminum or cast iron heads, all from the GM catalog.

A dynomometer is a great tool and not just for finding peak horsepower. With a dyno you can be sure the rebuilt engine is ready for the street, with the correct air/fuel ratio and no oil or water leaks.

spend, but how that amount will be spent. Some rodders take a reasonably good junk yard engine, give it a new paint job and then spend their entire budget on billet valve covers and polished intake manifolds. Others spend the same dollar amount and end up with a rebuilt and fortified engine that hides the rocker arms under stamped steel, painted valve covers.

Anyone who decides to do the rebuild and hop up at home needs to be careful with the parts they install. There is no best camshaft, carburetor or set of headers. Rather, there are a number of camshafts or carbs that will meet your goals when used with a particular group of complimentary components. What we're trying to say is that the cam and carb and all the rest needs to be chosen with an eye toward the total parts package. All these parts have to work together.

Q&A with Gary at Wheeler Engines

My Guest Editor in charge of engine building is Gary Schmidt, owner of Wheeler Racing Engines in Blaine, Minnesota. With more than twenty five years of engine building under his belt Gary has a good handle on how, and how much it costs, to build a strong street engine. Some street rod builders will opt for a simple used engine with chrome valve covers. The Q&A session with Gary is intended to provide insight and direction for all those builders who want more. Who believe that a little too much is almost enough.

When you take off the hood, the engine becomes a big part of the car, both mechanically and visually. How abaout a Hemi with two belt-driven blowers! Kelsy Petrykowski

Q: How much, in rough figures, to build a healthy small-block engine?

A: Assuming we already have a used small-block, thirty-five hundred dollars will build a good practical engine. That's a nice healthy street-able engine with 400 horsepower.

Q: If I budget that much

Old-time hot rods require just the right engine, in this case a 392 cubic inch Hemi with an adapted 440 water pump and hand-fabricated valve covers. Eric Aurand

money, now how do I spend it? How much goes for labor and how much for parts?

A: Of the total, twelve hundred dollars is labor for machining and assembly, and the rest is parts. The labor includes all the machine work; reconditioning the crank and rods, and balancing all the pieces. I recommend that these guys use an Edelbrock heads-and-manifold package.

We don't port small-block heads anymore, there are heads available for either Ford or Chevy that work well and flow lots of air. Edelbrock heads work up to 400 or 450 horses and then there are the Brodix and Dart heads if you want more. It doesn't make good sense to buy used factory heads and then spend seven hundred dollars to port the passages and fix them all up when for another two hundred dollars you can have new aftermarket heads. The new heads are a much better value.

Q: Can you outline a combination of parts that makes good power and still works as a street engine?

A: When my customer comes in and wants 400 horses I go to a 383 cubic inch engine (this is for Chevy small-block engines), essentially a stroked 350. Stroker cranks for the Chevy small-block are readily available and the extra cubes makes it a better street application because the horsepower and torque come more easily. I like the Edelbrock heads and manifold, a hydraulic cam with about 220 degrees of duration and a static compression ratio of 9.8 to one. We use cast pistons in most of these engines, the newer hypereutectic piston with a high silicone content.

An engine like this will run on pump gas and easily go over 400 for both horsepower and torque, at a relatively low RPM where most of us drive. I recommend a carburetor that's not too big, probably a 650 CFM, and a set of headers with 1-5/8 inch tubes about 30 to 32 inches long.

With a Ford engine it costs a little more and you have to work harder to get the same horsepower. There aren't the same kind of stroker cranks available, for example, so you're probably stuck with a stock displacement 351 cubic inch Windsor Ford. Otherwise the same basic combination of

You need to know exactly where the motor is going to sit before you build the car. In the case of the Steve Moal roadster seen in chapter 10, the full-size drawing shows both driver and engine location.

parts works on the Ford as well as it does on the Chevy engine.

Q: Do you often act as a partner to the person building the engine, in the sense that you advise them how much to spend and what parts to buy?

A: When a customer comes in I ask what his objectives are and what kind of budget he's decided on. Then I try to put together the best package of parts to achieve those objectives. Partly, I need to know if he's realistic in what he wants.

In certain cases I steer them away from some of what they want. For example, if they want really big horsepower numbers they might end up with peak power in an RPM range that isn't practical for the kind of driving they do. I have a "dream wheel" that shows rpm at a certain speed with different rearend ratios. That way I can make my customers understand that they might not have peak power until they're going 160 MPH. Some guys just want more horsepower than their buddy.

No polished billet here, just an early Corvette engine complete with two four-barrel carbs and a generator. Kurt Senescall

In those situations I have to talk them into a combination that provides the best horsepower and torque numbers in the RPM range where they drive.

Q: What do you think about fuel injection on street rod engines?

A: I think fuel injections is the future, for anything that's driven on the street. Because of the types of customers we have we don't see a lot of it in our shop. A beginner should stay as close to the stock engine as they can. The stock TPI gives lots of low end torque. The small intake runners and the design of the system provide lots of low end power. If you're working with a donor car and the car already has all of the necessary parts on it, you're already there.

If you really get into it and want to make modifications to an injected motor you need help

This 4.6 liter overhead cam Ford engine is slated to be installed in a'34 Ford at Jon Kosmoski's facility. Before buying the latest O.E.M. engine be sure you can find the correct computer and wiring harness.

from a person who understands fuel injection.

Q: How much more hassle and money is it to build a healthy street engine from something besides a Chevy or Ford small-block?

A: Obviously the cost of labor is about the same for most V-8s. The problem with those other engines is that most of them haven't been manufactured for over 20 years, so just finding the parts can be a hassle. Edelbrock has heads for the Pontiac engines and for the 455 cubic inch Oldsmobile, but not for the smaller Oldsmobile engines or the Buick Nailhead.

Q: You have an engine dyno at Wheeler. Explain how you use it and whether you recommend using it for all the engines you build?

A: I try to build the cost of the dyno runs into the package. That way when the engine goes out the door we all know everything works right, there are no leaks and the engine has good oil pressure. This provides us an opportunity to check the ignition and the jetting. With regard to jetting, we don't consider the dyno settings the final word. I recommend that the customer take it to a good chassis dyno and have the air-fuel ratio checked. We get it very close here, within one jet size. But due to the restriction in the street exhaust system, as compared to the open system we run on the dyno, the engine will probably get richer when it's installed in the car.

Q: What kind of mistakes do people make when they build a hot rod engine for the street?

A: If it's built at a reputable machine shop, the machining and the clearances are usually fine. Everybody understands those. The problems I see are in poor ignition and fuel systems. Problems in the tune up part of the engine. Even if we do a dyno run people might still have trouble when they get the engine in the car. Poor fuel delivery or an inadequate cooling system will cause the engine to detonate, which will pound out the bearings.

Fuel delivery is a big culprit. A customer might say, 'It ran good on the dyno but not in my car.' A lot of these cars don't have enough line volume or enough fuel pump.

The feed line to the fuel pump should be 3/8 inch with no kinks or fittings that create a restriction. The fuel pump should be a good mechanical pump designed with the large lines in mind. Something like an Edelbrock or Holley pump. No, you don't need an electric pump and you don't need two pumps. Just one good one. And there should be a fuel gauge in the pressure side of the system so you can spot a fuel problem before it ruins a new motor.

People don't understand that the new motor might need more cooling system than the old one did. If we build a 1000 horse motor people don't realize it takes a lot of radiator to cool that many horses, and many of these cars don't have enough frontal area anyway.

Whether it's a fuel or cooling system problem, people find out the hard way that these motors are not very forgiving when they're on the edge.

Q: Are complete crate motors a good deal?

A: The answer is yes, *if that motor provides the kind of power and performance the customer is looking for.* Our customers want more performance than you can get from a crate motor, and that's why they come to see us. But if you're looking for a good 330-horsepower 350, the crate engine is a good value. It's not a good starting point if you want to build a 500-horse engine, because you're going to throw too many of the new parts away.

If you're interested in a flathead, plenty of used engines and parts are available at the swap meets, and new parts - from aluminum heads to modern water pumps - show up in the magazines every month. Kurt Senescall

Chapter Seven

Transmissions & Rearends

What to Buy, How to Install

TRANSMISSIONS

When it comes time to choose the transmission for that new hot rod project, some builders simply use the transmission that came with the used engine. If, however, you don't have an engine and transmission assembly, or you want more than the standard TH350 that came behind the small-block you purchased, the choice may not be that easy.

First, decide if you want to install a clutch pedal to the left of the brake pedal Though it's often easier to

As evidenced by this Deuce frame and five speed transmission being assembled at Metal Fab, there is a new trend to install standard four and five speed transmissions instead of the tried and true automatic. The most intimidating part, installing the clutch linkage, might not be as hard as you think.

install an automatic, stick transmissions seem to be gaining in popularity and they do add a certain hot-rod flavor to your favorite car.

The clutch linkage is often the daunting part of adding a stick shift, though that's been made easier through the use of hydraulic clutch assemblies. No longer do you have to figure out pivot points for the under-hood linkage and how to mount a clutch pedal and pivot assembly. What you need is one clutch pedal, one master cylinder, one hydraulic line and one clutch slave cylinder. The slave cylinder itself can be mounted to the outside of the bell housing, or internally on the transmission's front bearing retainer. The internal slave cylinders, often called hydraulic throwout bearings, save space though you have to pull the transmission to service the slave cylinder assembly.

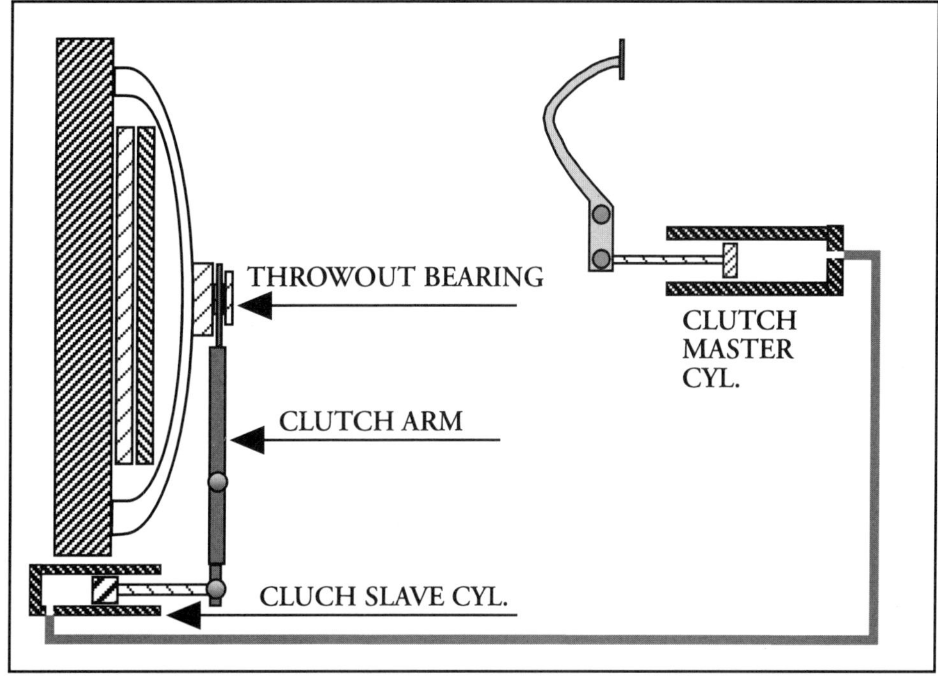

A hydraulic clutch simply replaces the mechanical cable or linkage with two cylinders and a hydraulic line.

Wilwood and others make clutch pedal assemblies that include the master cylinder and the pivot mechanism. Imported cars have used hydraulic clutch linkages for over twenty years, and now many American cars and trucks use hydraulic linkage as well. If you're installing a Ford engine and transmission a stock bell housing and slave cylinder, plumbed to an aftermarket master cylinder, might be the easiest way to proceed. For more on clutch linkage options check out the Q&A with Roy Brizio in Chapter 4. Note also that both the Hands-On cars in Chapters 10 and 11 are built using standard transmissions.

Automatic transmissions

With the introduction of four-speed automatics with lock-up torque converters, some of which are totally computerized, your list of possible automatic transmission options is longer than ever before. Though plenty of owners will opt for the default setting, a TH350, you'd be foolish to ignore the possible use of a more modern overdrive transmission. Advantages include a nice deep first gear for a good launch, combined with an overdrive fourth gear for quiet, low RPM highway cruising. The downside to

Combining the slave cylinder with the throwout bearing makes for a very compact assembly.

Though it doesn't offer an overdrive gear, the venerable TH350 remains a durable and popular automatic transmission.

the overdrive transmissions includes the added cost, and in some cases, complexity. In the case of the 700 R4 from GM, the transmission is bigger and longer than a 350, meaning you must notch or modify the cross-member in a typical fat-Ford frame.

Choosing an Automatic Transmission

For help in choosing the right automatic we offer the following look at the most popular transmissions, with comments from another guest editor at the end of the section.

GM

The Turbo Hydro 350 automatic transmission is still the most popular of the automatic boxes manufactured by GM. Built from 1969 to 1979 and from 1980 to 1986 in a lock-up version, there are millions of these transmissions out there in the junk yards and at the Sunday swap meets. Advantages include the relatively small size and short length and the ready availability of parts. The earlier, pre-lock-up, transmissions are the more desirable units. Most of these housings are the same with the exception of some 4X4 housings and some reinforced housings used in trucks.

Any difference in length on a TH350 is in the tail-shaft housing. The most common housings measure either six or nine inches long. Total length with the short housing is just over twenty-eight inches from the edge of the bell housing to the end of the tail-shaft housing. The rear mount bolts to the main transmission case, not the tail-shaft housing, so the location of the rear mount is not affected by your choice of tail-shaft housings.

There is also a difference in the diameter of the speedometer drive housings.

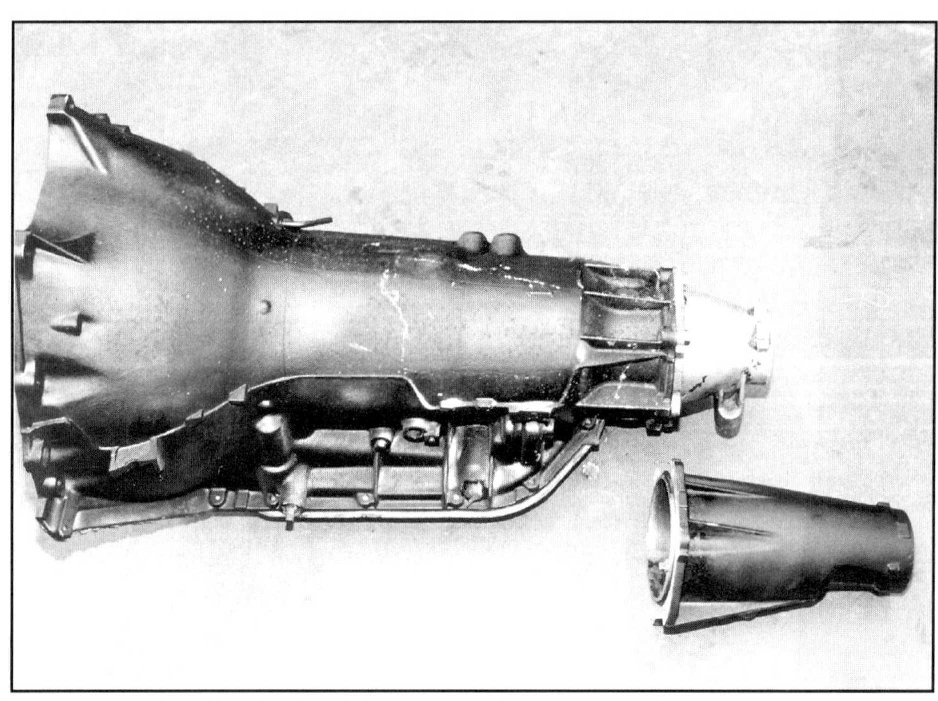

Seen here with the non-Chevrolet bell housing pattern, the TH400 is a very durable transmission, capable of handling all but the most severe abuse. Though they're hard to find used, the short tailshaft housing makes for a much shorter transmission.

While Chevrolet transmissions carried a speedo housing that was only about one inch in diameter most of those used in Buicks, Oldsmobiles and Pontiacs have a much larger hole and speedometer-gear housing. Installation of a 350 transmission will require a shift linkage, a vacuum line to the modulator and a kickdown cable.

The Turbo Hydro 400 might be called the 350's big, brutish brother. Installed behind many high horsepower engines and in many GM trucks, this transmission is slightly larger in diameter, heavier and longer by one inch, than the 350 transmission. For high horsepower applications, or the individual with a very heavy foot, the TH400 might be the right answer. Like the smaller 350, you need shift linkage and a vacuum line to the vacuum modulator. Instead of a kickdown cable the 400 uses an electric switch to activate passing gear.

A special, switch-pitch, TH400 was offered for sale from 1965 to 1967. This interesting alternative used a stator (the vaned unit positioned between the drive and driven members of the torque converter) with pivoting vanes. In essence, the angle of the vanes could be changed from a *high stall speed* for acceleration to a *low stall speed* for efficient highway cruising.

The switch-pitch 400 makes a good street rod transmission, though they might be hard to find. These transmissions were offered in Buicks, Oldsmobile, Cadillacs and Rolls Royces during the two year period. If you need one of these hard-to-find transmissions a specialty automatic transmission shop is the best place to look

The GM overdrive transmissions include the well-known 700 R4, a true four speed transmission with a lock up torque converter. This transmission allows you to combine a low first-gear ratio of 3.06:1 with an overdrive fourth gear of .70:1, so you can have a good launch and low revs in the highway, all in one package.

As mentioned by Greg Ducato in his Q&A, a good R4 can handle substantial amounts of torque and horsepower. R4s come with two bell-housing bolt patterns: 2.8 V6 and four cylinder cars used a bell housing with a slightly different (sometimes called *metric*) bolt pattern while the 4.3 V6 and all small blocks have the correct bolt pattern. To check, make sure the two upper mounting holes measure 8.25 inches from center to center.

R4s built after 1987 are superior to earlier units (these have pressure ports on the side next to the servo, while the earlier models do not) though the earliest R4s can be upgraded with the latest parts. If the trans-

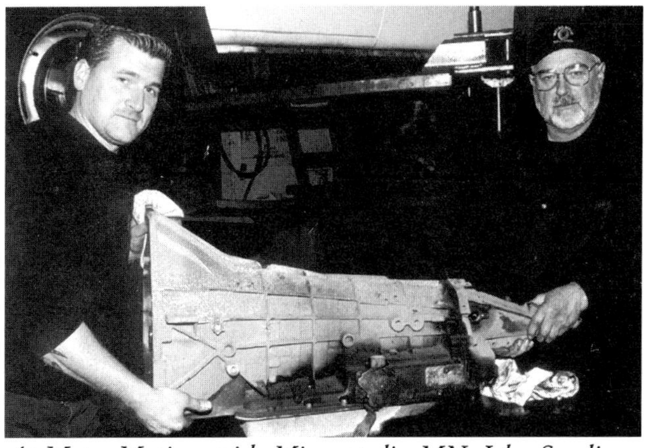

At Metro Matic outside Minneapolis, MN, John Sevelius and his son Randy do a mixture of hot rod and late model automatic repair. The huge EAOD they hold in their hands is the electronic version of the Ford AOD - A transmission they feel is too large for most hot rod applications.

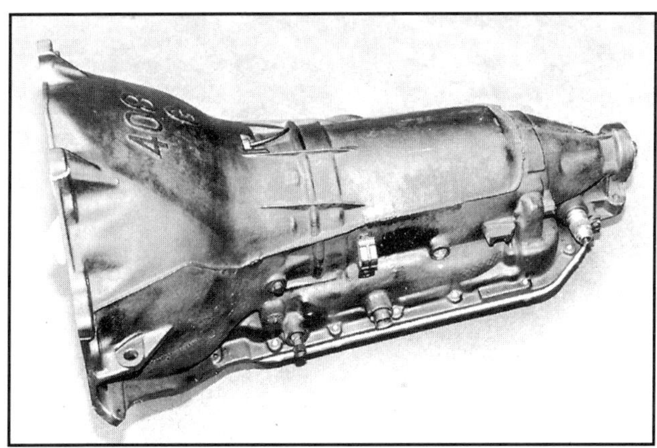

The 200 4R is GM's smaller overdrive automatic. Note the integral tailshaft housing.

Originally built to handle only 300 foot pounds of torque, the 700 R4 can now be built to handle even serious big-block power.

From bottom to top, a look at the relative size of three GM trannys: a TH350, TH400 (with long tailshaft housing) and 700 R4.

mission you're using is four or five years old, it's probably a good idea to have it overhauled before installation anyway. If it's an early model the transmission shop can upgrade it to the latest specifications and install heavy duty clutches at the same time.

GM introduced the 200 4R overdrive transmission in the early 1980s. These are not modified 350s and are not part of the 700 R4 line. As mentioned by Greg, these are perfectly good, four-speed transmissions with most of the advantages of an R4 without the bulk and expense. The 200 4R has the advantage (shared with the 350) of smaller size which means it requires little or no modification of the X-member when installed in most fat-Ford frames.

Ford

The three most popular Ford automatics include the C4, C6 and the newer AOD automatic. While the C4 is fine for small-blocks of modest power, the C6 is a much stronger transmission. Though it's an older design, parts for the C6 are still readily available. The AOD is used behind current Ford small-blocks including Mustang GTs. This is a true four-speed transmission though the lock-up torque converter is controlled mechanically.

Chrysler

Chrysler transmissions include the very durable Torqueflite design, produced in two versions: the 904 and the 727. The 727 is by far the superior unit from a standpoint of strength though a 904 will work just fine with small-blocks of reasonable power. In size the two transmissions are very similar, with the 727 being a bit thicker through the center of the housing. The 727 can be built to street-hemi specs and more. In addition, there are two overdrive versions of the old Torqueflite, the A500 and A518, which correspond to the 904 and 727 respectively.

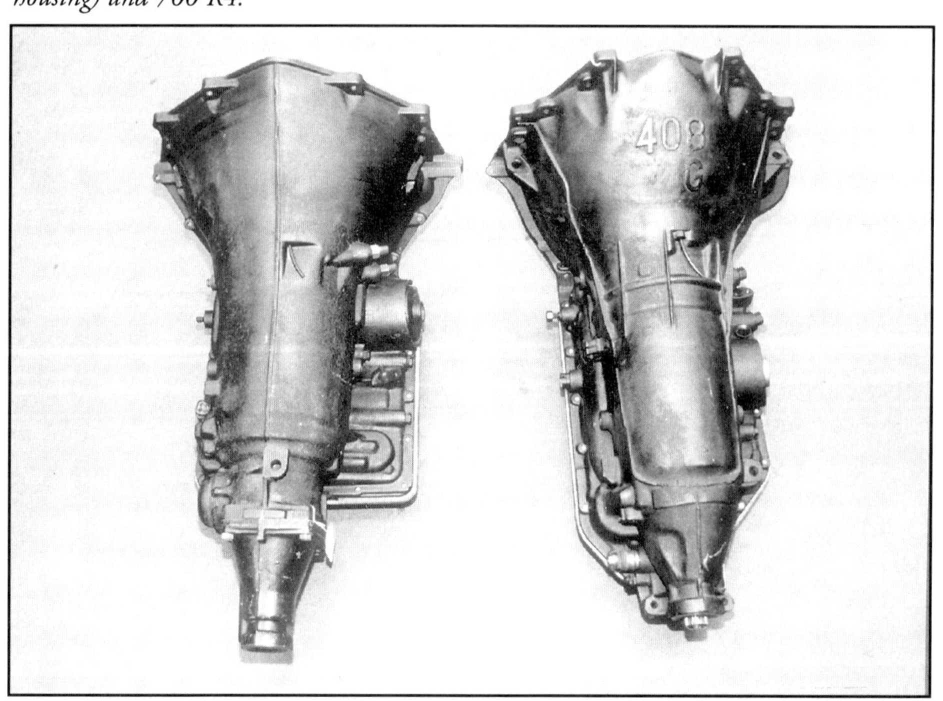

The 700 R4 on the left is physically bigger than the 200 4R on the right, which can cause installation troubles. The biggest problem isn't the length (which is exaggerated here) but the way the oil pan and housing jut out on the right side.

Tranny tips

Before installation it's an excellent idea to spend the bucks and have a used transmission overhauled, unless you're really sure the unit in question is in excellent condition. At a minimum, change the fluid and filter in the transmission before you run the car the first time. There is no way to check the clutch in a sealed lock-up torque converter. You can ensure the clutch is good only by buying a rebuilt torque converter.

Transmission failure is generally the result of too much heat. Trailer towing and drag racing (sanctioned or otherwise) can cook that nice rebuilt transmission in short order. To prevent the problem install a transmission cooler, which will also make the radiator's primary job of cooling the engine a bit easier.

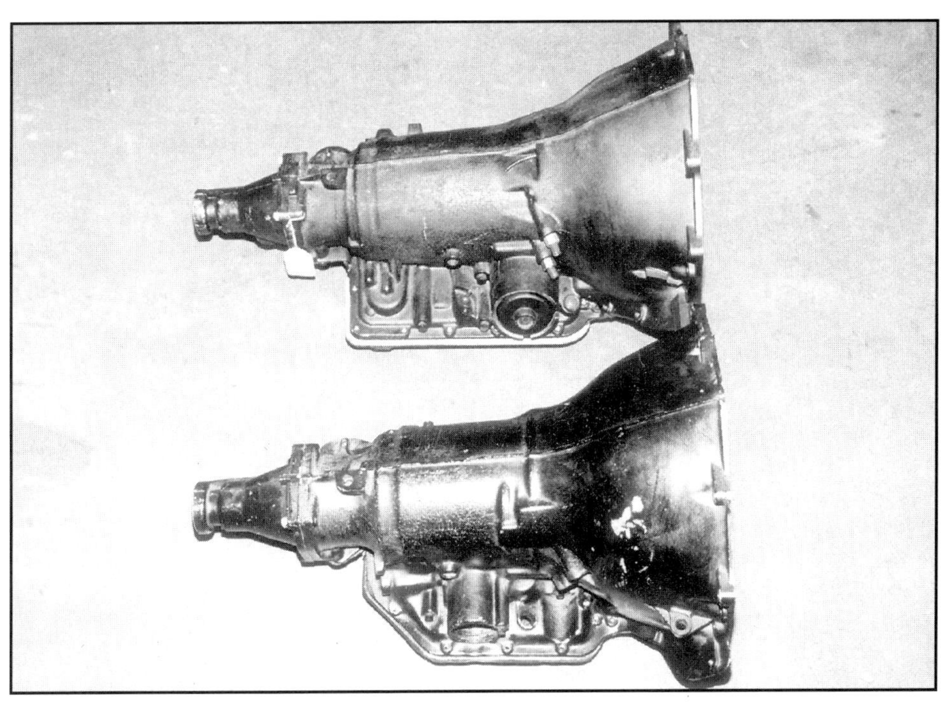

Because they are nearly the same length the 700 R4 (top) will go in where a TH350 came out, if it will clear the X-member.

Q&A: Greg Ducato

For help deciphering which transmission, from among the many available, is the best candidate for our particular street rod I called Greg Ducato, owner of Phoenix Transmission Products. When I asked Greg what made Phoenix different from all the other transmission shops around the country he explained that at Phoenix, "We specialize in building transmissions one at a time for specialty cars and street rods. It's like buying a tailored suit, you don't want one off the rack, you want one tailored to your car and driving style. We can adjust shift points, shift firmness and speedometer drives. That way the customer gets exactly what he wants. After 16 years in business it's become pretty obvious to me that all the engine work in the world won't help if you can't get the power to the rear wheels."

Another look at the 700 R4 alongside a TH350 (on the right). Note how the 350 is contoured on the right side, almost like it was designed to clear an early Ford X-member.

Q: There are a number of automatic transmissions available now from each of the big three. In the GM line there are at least four different automatics that are commonly used in

AUTOMATIC TRANSMISSION GEAR-RATIO CHART

	First	Top Gear	Lock-up
TH350	2.52	1.00	N/A
TH400	2.48	1.00	N/A
700 R4	3.06	.70	Electrical
200 4R	2.75	.68	Electrical
AOD	2.48	.68	Mechanical
A 500	2.48	.69	Electrical
A 518	2.48	.69	Electrical

Don't just assume you need a certain automatic transmission. Give it a little study because the features are very different and have a huge impact on the car.

street rods. How does a builder decide which of these is the best one to run in a new hot rod?

A: You need to ask yourself what are you going to do with the car. Will it be a competition car or are you going cruising. You have to base your decision on the power level of the engine, the expected driving and any anticipated abuse. Price is a consideration as well.

Q: In the GM line, can you list some of the advantages and disadvantages of the TH350, TH400 and the overdrive transmissions?

A: A TH350 is a good transmission for a daily driver kind of car built on a budget. A 350 will match up well to a 2.78 or 3.00:1 rear gear and a mild small-block. But if the engine is a big-block or has a lot of power, like over 500 horses and you intend to run it hard, then consider the TH400.

The C-6, available in either a big-block or small-block bell housing, is the strongest of Ford's non-overdrive transmissions. Though it hasn't been made for some time, parts and complete transmissions are readily available.

If you intend to run a steep rear gear then you need an overdrive like the 200 4R or 700 R4. The 200 4R works OK for a cruiser, maybe you're going to take the wife and kids out on weekends and the car has a 300 horse crate motor under the hood. Then the 200 4R is great. It uses less horsepower, comes with a 2.75:1 low gear and a .68:1 overdrive. The 200 is easy to mount where a 350 came out because it's the same length. It's also the least expensive of the overdrive transmissions. We sell quite a few of them. The 200 might not be the one to use with a real high horsepower engine and a heavy car, but it's a good proven transmission. We see them with 150,000 miles on the clock that have never had any service.

If you want an overdrive transmission and you've got more power, say 400 horses or more, then you might want to consider the 700 R4.

This transmission can take a lot more power than the 200 4R and can live behind big-blocks in most street rods. Compared to a TH350, the 700 R4 is just as good in terms of longevity and it can do a lot more things.

Q: What are the mistakes people make when they choose and install a transmission in their street rod?

A: You need to really know the direction you're going with the car and base your decisions accordingly. Will this be a race car or a cruiser? A lot of this has to do with the rearend ratio. If you have a 2.7 or 3.0:1 rearend and lots of horses, you don't use an overdrive. But if it's a 4:11:1 gear then don't use a 350 or 400 transmission. If you do, you can't drive it any distance because the engine is spinning so fast on the highway.

So many people call me on the phone and they have this great engine with a blower and gobs of power and they're asking me about running a R4 because they want to get some mileage from the car. And I think, 'why waste your money on a R4 in an attempt to get mileage when the original intent of the car was to build a real hot rod?'

There's so much stuff out there that it's easy to get distracted by the options and lose the focus you had for the car. People read about this guy or that guy and his car in a magazine, but the magazines don't do a good job of really describing the technical side of the car and they also don't tell you if the builder has had any trouble with the car or had to change the rearend ratio to make the engine/transmission package work correctly. We build transmissions per a custom order. The tranny is matched to the owner's package of engine, rearend ratio and expected use so it works like it's supposed to.

Q: How about a quick look at automatic transmis-

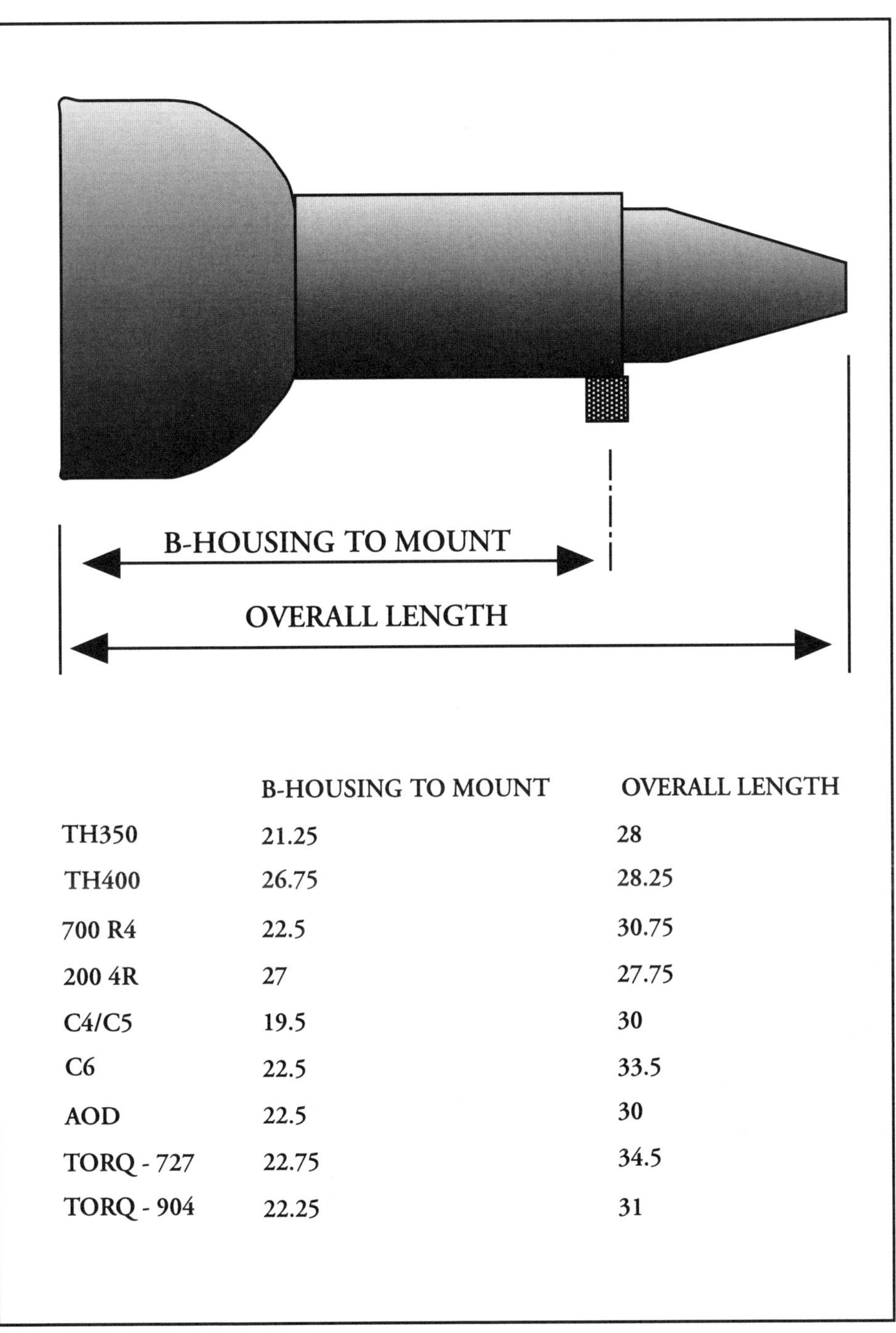

	B-HOUSING TO MOUNT	OVERALL LENGTH
TH350	21.25	28
TH400	26.75	28.25
700 R4	22.5	30.75
200 4R	27	27.75
C4/C5	19.5	30
C6	22.5	33.5
AOD	22.5	30
TORQ - 727	22.75	34.5
TORQ - 904	22.25	31

Length and mount position are only one of the factors that must be considered when trying to choose the right transmission for your next project.

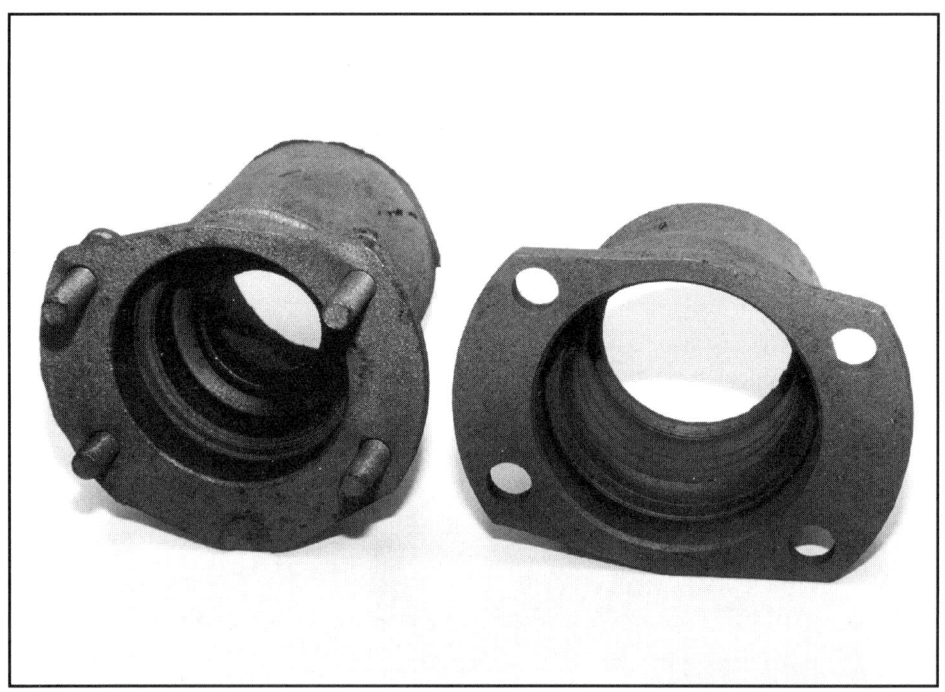

As mentioned in the text, the Ford nine-inch rearend housing comes in either a big or small-bearing version. The big-bearing housing is preferable because it also comes with the larger diameter axles.

sion options for Fords and Mopars?

A: In the Ford line there's the basic three-speed transmissions, the C4 and C6, and the AOD or overdrive automatic. Basically the AOD fits small blocks and has the same pluses and minuses as the GM choices and the same price parameters as well.

In the Mopar line there's the older 904 and 727, and overdrive A500 A518. The A500 and A518 are the overdrive equivalents to the 904 and 727. The only difference is that both the 500 and 518 come with small-block bell housings only. The 518 can be used behind a big block with an adapter. Like the 727, the 518 is a super, heavy duty transmission, far huskier and better in terms of the integrity of design than anything else on the market. It's more expensive too, it's in the fifteen hundred dollar range.

Q: What about using the modern overdrive automatics in cars that don't have computerized engines?

A: With an GM overdrive, a 700 R4 or 200 4R, there's only one wire that controls the lock up for the converter. And we wire those up internally so you don't need a computer. The Ford AOD uses mechanical hook up of the torque converter so there is no black box. We convert those to non-lock up transmissions, so they don't feel so clunky. For the more truly computerized automatics there are stand-alone black boxes available so you can run them in a car that doesn't have the original factory computer.

Q: Any final advice on building a street rod and installing an automatic transmission?

A: Don't get carried away

Pre-'84 Corvette rearend housings are available in either cast iron or aluminum. Though the cast iron might be stronger the aluminum is light and of course it can be polished to a bright shine.

with power, these are light cars to start with. And you never run over 5000 rpm, so don't build a motor that doesn't come alive until it hits 4500 RPM. Stick with reliable set ups instead of the exotic stuff. Keep sight of the 'big picture.' Tire size and gear ratio are just as important as the color you paint the car, but a lot of people miss that. Too many features in the magazines don't explain the mechanical part of the car so nobody understands the importance of tire size and rearend choices and which transmission to run.

Though not as sexy as a polished independent rearend, a simple solid axle with chrome four-bars or ladder bars, and chrome or polished panhard bar and shocks makes for a very neat installation.

If they have a transmission builder in mind, ask him as many questions as possible. I find that transmission and engine builders are happy to give the customer as much information as possible because it helps us if the customer is informed. People shouldn't be afraid to ask what seems like a stupid question. If they don't ask the question up front, it may come back to haunt them later.

And you've got to build the car for yourself, don't built it to please the judges or your buddies.

Rearend options

It seems that everyone wants a nine-inch Ford rearend. Yes, they're strong and stout and a good logical choice but there are some other options out there. (For more on nine-inch rearends see the side-bar in this chapter.)

A short list of options would start with the eight-inch Ford rearend used in many mid-size cars. Though not as durable as the nine-inch, this is a perfectly good rearend and it is a third-member design. Narrowing a rearend adds expense to the building project.

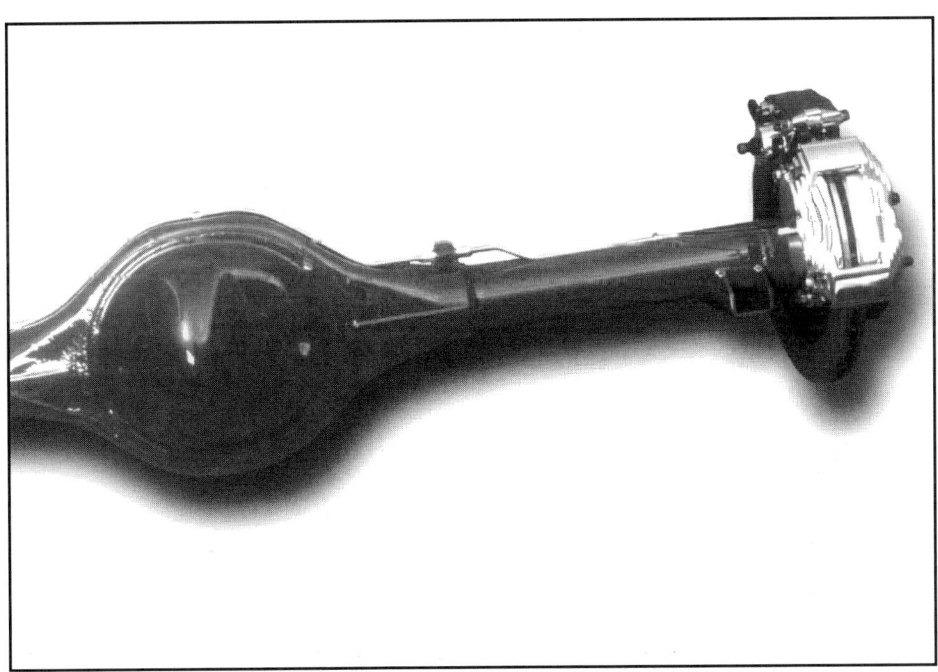

If you don't want to scrounge the junk yards, complete nine-inch housings and center sections, with brakes and dimensions tailored to your car, are only a phone call away. Dutchman.

This very heavy duty Ford nine-inch features a reinforced housing, aftermarket housing ends, heavy duty center section and "unbreakable" axles.

The nice thing about the smaller Ford rearend is the fact that some are only fifty-six inches wide, flange to flange, which means they fit many Deuces and later street rods. You might even find one with factory disc brakes. Ford also makes an 8.8 inch rearend, standard equipment on Mustang GTs. These units are also available brand new, with Traction-Lok, from Ford Motorsports SVO.

On the bow-tie side of the aisle the smaller, ten-bolt rearend used in many Novas and similar products is popular, though most of these rearends measure over sixty inches from flange to flange. These rearends are hard to narrow as well, because the axle is tapered behind the splined area. A better choice is the rearend assembly from a mid-1960s Nova measuring only fifty seven inches from flange to flange. This convenient size can be fairly difficult to find.

The 12-bolt GM housing is less popular here because it is not a third-member design and the size typically means these rearends must be narrowed.

The eight and three-quarter inch Mopar rearend is a good durable unit and it is a third-member design, though it's hard to cut the axles due to their shape. Narrowing a Mopar rearend will usually mean buying shorter axles. Some relatively narrow assemblies were used in Dodge Darts and Plymouth Dusters equipped with the 340 or 383ci engines. These are probably a good choice for a street rod, though again they can be hard to locate.

Early style, cast iron Corvette housings are still available, quite durable and can be matched up to a larger variety of gear sets than the sexier aluminum housings.

Mounting the rearend

Before mounting the rearend you need to know the position of the axle and suspension. While you're doing all this planning remember that the wheelbase measurement isn't set in stone. Most of the Ford products, and many GM cars, had the rear tire positioned near the front of the fender opening. You may want to put the tire farther back in the opening for a more balanced appearance.

Spend time mocking up the rear axle assembly with wheels and tires to make sure the finished product will sit where you think it should. Ask the manufacturer of the rear suspension kit where their parts and instructions will place the rear wheel and tire relative to the fender opening.

Where to mount the motor

The position of the engine in the car is obviously important, but for street rods and hot rods it's doubly important. Some of these cars have very short engine compartments. Also, many builders place the engine very low in the chassis. While this may create a nice low center of gravity it puts the water pump snout pretty low in relation to the radiator. The water pump snout is of course the mounting point for the belt-driven fan. If the water pump, and thus the fan, end up pretty low in the frame then you can only run a very small diameter fan.

A spool eliminates the carrier and spider gears to provide you with a locked rearend.

For cars with engines set low in the frame there are electric fans or special water pump risers that raise the water pump hub. Most professional builders (I'm sure to start an argument here) prefer to pay particular attention to engine position during the engine and frame mock-up. You have to make sure the engine is high enough so the fan ends up in a good location; and back far enough to make room for the fan, radiator, condenser and a shroud.

Like everything else, mounting of the engine and transmission is one of those building stages that requires time and patience. Consider that engine position affects

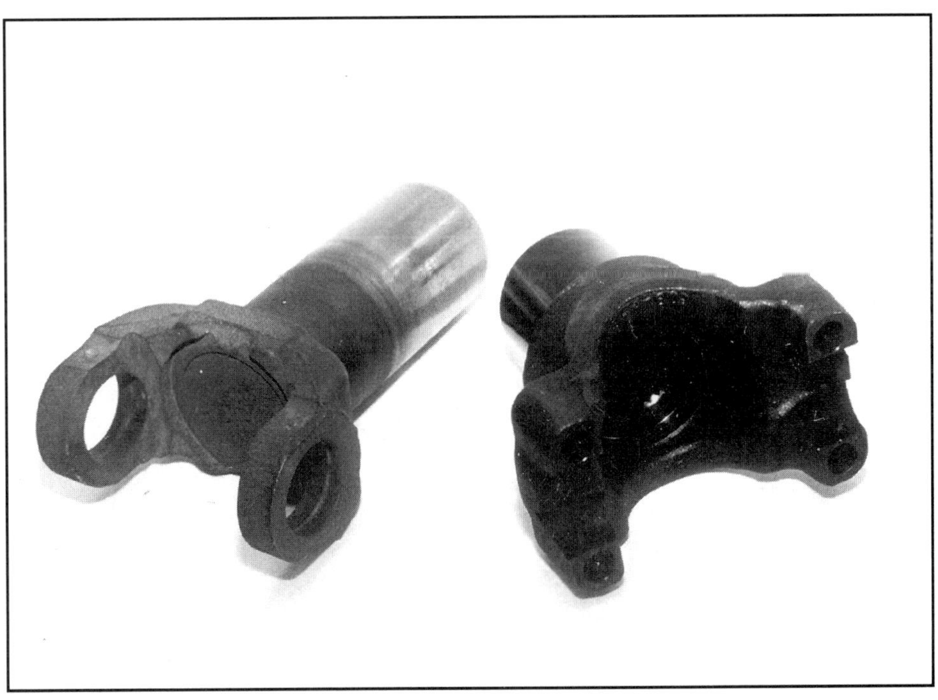

Not all yokes are created equal (this is no laughing matter). Even though the spline count might be the same, heavy duty yokes allow you to use larger, heavy duty U-joints.

Ford Nine-Inch Rearend

Number nine

The nine-inch Ford rearend is very popular and the reasons are simple: It's strong, it's relatively easy to narrow and it uses the "pumpkin" design with a removable center section.

The nine-inch Ford gets its strength both from a strong housing and good stout gears with good contact between those gears. The Ford housing (there are different models available) is generally stronger than either the 8-3/4 inch Mopar or the 12-bolt GM housings.

The first nine-inch Ford rearends were used in passenger car and trucks as far back as 1957. Passenger cars used the nine-inch only until 1973, the pickup trucks used them until 1984, and full-size Ford vans used the nine-inch though 1987.

If there is a downside to the nine-inch Ford design it might include the price and the availability of used positraction units. Though thousands and thousands of these rearends were manufactured, demand by street rodders, drag racers, off-roaders and anyone else interested in a really tough rearend means that prices aren't always cheap. If you want positraction with your rearend, the availability goes down and the price, of course, goes up.

"Sometimes when first-time builders call and ask about buying a nine-inch rearend complete, they get sticker shock," explains Todd from Dutchman Motorsports, a company that specializes in high performance rearends. "A complete housing with posi center section and disc brakes is over two thousand dollars. But the guy who doesn't mind going out and getting his own rearend, cleaning it up and then bringing it to us to have narrowed, he can get by for a whole lot less."

The problem with narrowing almost any rearend is the fact that most axles have a smaller diameter just inboard from the splines. So when you cut off the splines the axle shaft at that point isn't large enough to re-spline. In these situations you have to buy new axles. Ford axles have a short small-diameter section so it's more likely you can shorten and re-spline an existing axle.

Though it looks pretty ugly, this nine-inch housing is a "good"one from a pick up truck. Good in this case means it's long enough that the axles can be shortened and re-splined, and the housing has the large diameter bearings.

The third-member style design means that you can drop out the center section for work or maintenance, or to install another center section with a different ratio, fairly easily.

Variations

The Ford nine-inch was offered in a variety of housings and with a variety of axle and bearing combinations. First, beware of imitations. Ford manufactured a 9-3/8 inch rearend, visually quite similar to the nine-inch. The trouble is there are few parts available for this odd-ball rearend. You know the rearend is the larger model if you can place a socket straight onto the housing stud at the 7 o'clock posi-

Ford Nine-Inch Rearend

If you can't find a nine-inch at the junk yard, new ones are available from a host of sources. The example seen here goes beyond heavy duty. Features include a new nodular iron housing, a billet pinion-bearing support assembly and the monster heavy duty yoke.

Inside it's more of the same: Billet spanners locate a Lenko 35 spine "detroit locker" center section with the biggest ring and pinion that will fit a nine-inch housing.

tion. The other visual difference is the top, horizontal rib on the housing. If the rib turns sharply down at one end, it's the nine and three-eighths inch rearend. (Note, a good shop can easily modify the larger housing to accept a nine-inch center section.)

The nine-inch center-sections came in two basic styles with one variant. More than ninety per cent of the housings seen at swap meets have a single vertical rib at the top of the housing. A few are the stronger, two-rib design that was used during the first few years of production. Rarely you will see a nodular housing, made from extremely strong nodular iron with a large N cast into the housing itself.

The nine-inch rearends came with different axle diameters, different axle spline counts and different size wheel bearings. Axles came in 28 and 31 splines, in two shaft diameters. The larger diameter axle uses a larger wheel bearing. Thus the larger axle/bearing combination is not interchangeable with a small diameter/small bearing housings. A 31-spline axle is always the larger diameter shaft, but a 28-spline axle can be either the large or small diameter axle shaft. (Confused yet?)

Large diameter axle shafts with the larger bearings have an axle housing I.D. (at the wheel end) of 3.150 inches. Small diameter axles with smaller bearings (the smaller diameter axles are always 28 splines) have a

Ford Nine-Inch Rearend

housing I.D. of 2.89 inches. In terms of relative strength, the 31-spline design is thirty-five percent stronger (torsionally) than a 28-spline shaft cut on the same diameter axle. These axles are stronger due to their larger diameter at the spline and the increased number of splines.

Besides the two different sizes of axle bearings, there are different styles of bearings as well. One of the most common is the tapered roller bearing. This bearing is lubricated by rearend lube and features a seal on the outside of the bearing. New bearings of this style are easy to find and purchase. Special sealed bearings that will replace the tapered roller bearing are also available. Single and double-row ball bearings were used by the factory as well. Some of these are sealed bearings and use a separate seal on the inside of the axle. Some trucks used a 31 spline axles and an odd ball double-row ball bearing that's hard to find. A special wheel bearing is available from Summer Brothers that allows you to combine the large diameter axles with a housing designed for the small diameter axle/bearing combination.

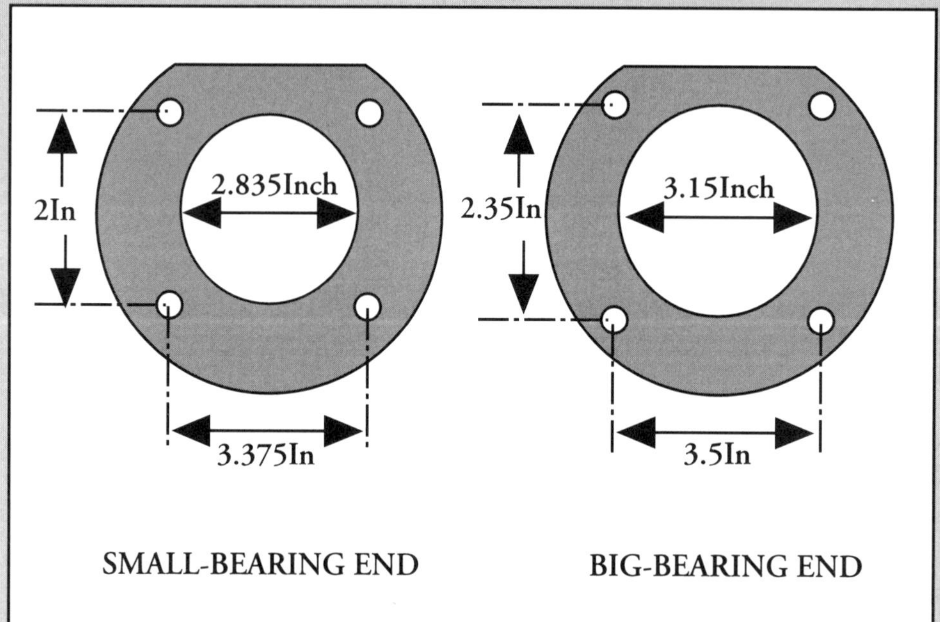

With the measurements provided here and a tape measure you can tell exaclty what it is you're buying when you go looking for that Ford nine-inch housing.

When it comes time to buy a nine-inch housing you either want a hard-to-find narrow rearend that came in some Granadas and Lincoln Versailles, or a very wide housing from a truck or boat-anchor station wagon. A truck rearend will provide you with a large, heavy housing and axles that are both large in diameter and long. The longer axles will have enough length so you can cut off all the tapered section and still have enough length to re-spline. Larger diameter axles are probably 31 splines. If not they can be cut to 31 splines when the ends are re-splined (but then of course the spider gears or the complete third member must be changed as well).

Positraction for the nine-inch

A variety of what we call "Posi" units are available for the Ford nine-inch design. The Equa-Loc and Traction-Loc are similar in design. Both were offered in two and four pinion designs, with four pinions being harder to find and much stronger. Both designs use clutch discs to control differential action. The Traction-Loc is the better of the two designs and is available in both 28 and 31 splines. The Equa-Loc is available only for 28-spline axles.

More durable and expensive is the Detroit Locker. This is a true locking differential with both wheels positively driven whenever the car is moving straight down the road. Meant for serious performance use, individuals who use them on the street report that they can feel them "work" on a corner and that they make a certain amount of noise. New and used center sections, with and without various forms of positraction, are available from companies like Currie, Dutchman and others.

everything else: the location of the radiator, shroud, fan, and condenser. Take your time, do some planning first. Because if you do it wrong it's a lot of work to correct.

Sketches for the car should include the position of the engine in three dimensions. You should know where the engine will sit, and where that puts all the other components. Positioned closer to the firewall is generally preferable. By setting the engine back as far as possible, within reason, more of the weight is shifted to the rear of the car. You also create more room for a decent multi-blade fan, a nice thick radiator and a good condenser. (Check out Steve Moal's comments on engine location in Chapter Ten.)

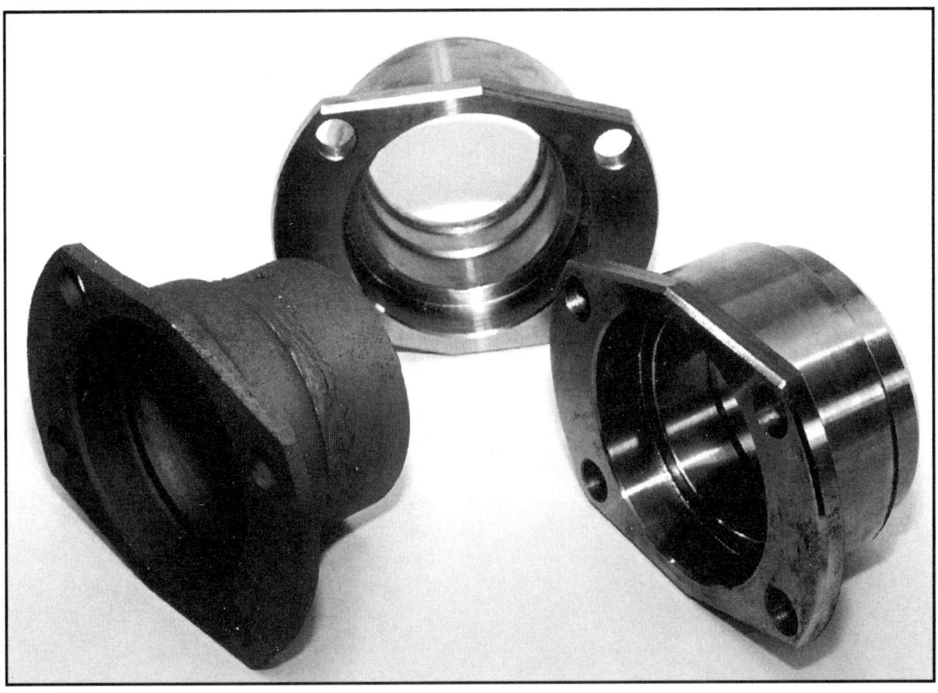

Billet housing ends, available in either small or big-bearing sizes, are often used in place of the stock Ford pieces when a nine-inch is narrowed.

Jack Chisenhall from Vintage Air encourages builders to mock-up the engine with a water pump and fan bolted in place and the radiator and front sheet metal set in place on the frame. As Jack is fond of saying, "Sometimes if they would just take a little notch out of the firewall they could move the engine back a few inches - and create enough room for the right radiator, a shroud and condenser. Enough room so the final product is a car the owner can drive anywhere. One with good air conditioning that doesn't overheat."

You should center the engine between the two side rails. To keep the center of gravity low you might want to get the engine pretty low in the frame. Remember, however, that the bottom of the oil pan is one of the most vulnerable parts of your street rod. Five inches is a good minimum figure for the distance between the oil pan and the ground.

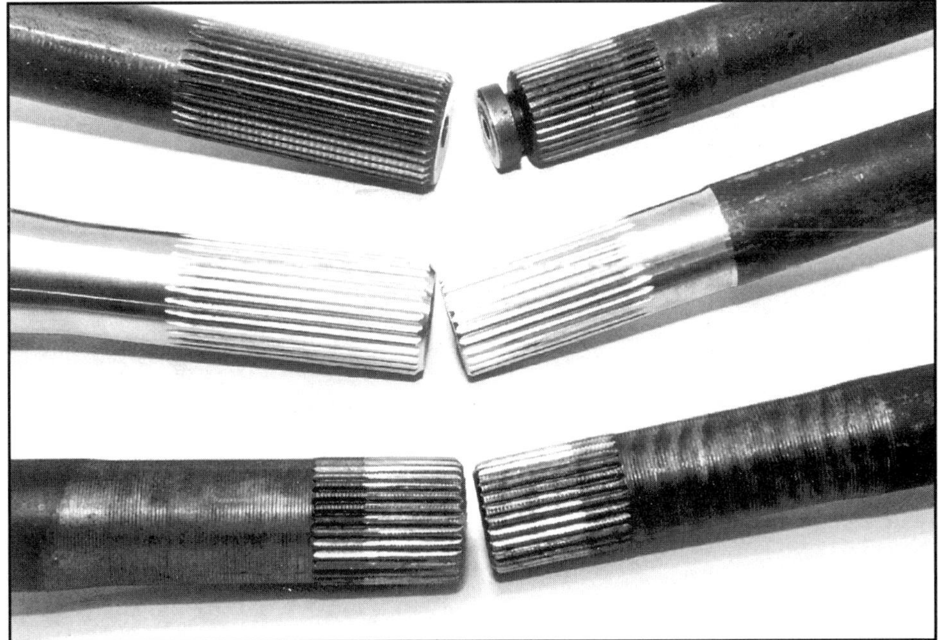

A plethora of axle ends, starting on the bottom with nine inch, big and small-diameter, 31 and 28-spline stock axles. Note the slightly smaller diameter of the axle just behind the splines. In the middle are two, shortened and re-splined, large and small diameter Ford axles. On the top left is a 35-spline aftermarket axle and across from that a stock GM axle with the groove for a "C" clip.

For extreme heavby duty the aftermarket makes larger diameter nine-inch axles in 35 and 40 aplines.

You also need the engine to be high enough in the frame so that the belt-driven fan is in the center of the radiator - which is why you need to do the mock-up with the fan bolted on in the first place. Know the location of the firewall if you are setting the engine and transmission into a bare frame. Most shops consider the correct angle for the engine to be lower at the back by one or two degrees. You can put a level across the carburetor mounting base to ensure that the engine is mounted so that the carburetor is level.

Most modern engines are bolted to the frame with three engine mounts, two in front and one on the rear of the transmission. The front mounts handle most of the weight and the enormous torque of a strong V-8. Be sure the mounts you use are strong enough for the job at hand. We should note here that engines from the early '50s often used a plate on the front of the engine, tied to either frame rail, and two more mounts on either side of the bell housing. If you decide to run an early engine, take time to figure out the mounting system, it can be tough to convert these early engines to the current three-mount set up.

Different size and style bearings are available for nine-inch housings. Some are sealed while others are lubricated with rearend lube. In addition, there are special bearings that make it possible to run large diameter axles in a small diameter housing.

Positioning the engine and transmission in the frame and mounting it correctly is made much easier with a small support rack. This can be a simple affair, just strong enough to hold the engine and transmission up off the table or frame jig. If you make it a little too short, shims can be used to do the final positioning for engine height and angle. Without the engine support stand you're trying to position an engine while it dangles on the end of a chain. Some shops run a piece of square tubing across the frame rails and then put two bolts into two of the holes for the front timing cover or water pump. This helps to position the engine and also ensures that it is level from side to side.

Drive-line angles

The following comments on drive line installation and drive shaft angles are compiled from the Boyd Coddington how-to books.

Most U-joints are designed to work in a range of five to ten degrees, ideally closer to five than ten. Under acceleration the angles of the U-joints (especially the rear U-joint) will change. The amount of this change will vary according to the design of the rear suspension. Trouble comes by using a U-joint and yoke assembly that doesn't allow much clearance between the U-joint cap and the yoke. If the angle gets fairly steep under acceleration or deceleration the cap may contact the yoke. This problem is sometimes mis-diagnosed as weak U-joints or too much horsepower.

The eight-inch Ford rearend is another good choice for hot rods. Though not quite as strong as a nine-inch it's a good third-member design, and positraction units are available.

When a drive shaft is built or modified (generally a job best left to a professional shop) the U-joints must be kept in phase. Drive shaft lengths are commonly figured with the front yoke moved back from the bottomed position three-quarters of an inch. Measure the shaft from the center of one cap to the center of the other. Shaft construction should include the cost of balancing the finished shaft.

When you mock-up the motor in the chassis it's a good idea to know exactly where the radiator (and firewall) are going to be. That way you can be sure there's enough room for a good radiator, and a fan to draw air across the core.

There are some very different opinions among street rod builders regarding drive

Use a shroud on your radiator - it will help to lower the coolant temperature and improve cooling efficiency.

shaft angles. One group feels that the pinion angle (the angle of the centerline of the pinion shaft as seen from a side view) should match the angle of the engine and transmission, while others insist that the two angles should be different by at least one or two degrees.

Jim Petrykowski, a man who builds everything from very short wheelbase Anglias to large and long, fat fendered Fords, feels that the only time the angles should be the same is in the case of short-wheelbase cars with gobs of horsepower. When setting up a dragster the driveline is setup to be in one straight line so all the parts are better able to transfer enormous amounts or horsepower and torque.

In the case of more common street cars and street rods in particular, the engine "centerline" is usually at a different height than the pinion. Jim likes to have a difference of one or two degrees between the angle of the engine and transmission, and the angle of the pinion (in side view).

As already mentioned, in the top view, the rearend often has an offset to one side. Jim usually leaves this offset intact, assuming that the dimensions of the finished rearend housing will place the driveshaft in the center of the driveshaft-tunnel. In the top view then, the centerline of the pinion and the engine are parallel though they may be offset slightly.

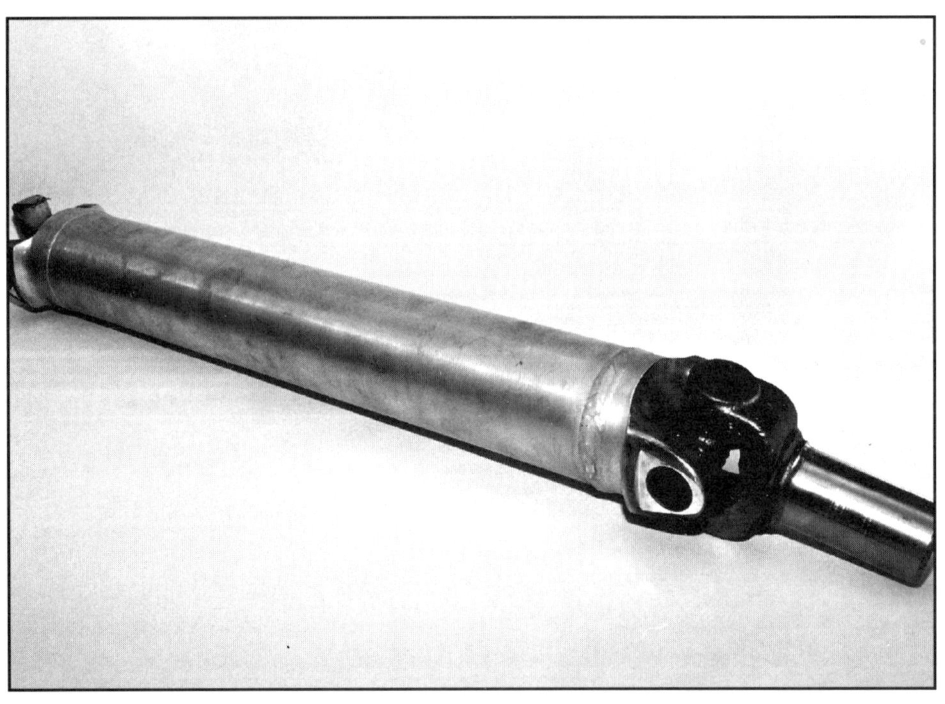

Aluminum driveshafts have become popular. Torsionally, the aluminum takes a lot of shock and the light weight means less reciprocating weight. Aluminum shafts are used in some O.E.M.s and the price of aftermarket shafts approaches that of steel.

In the side view, Jim prefers to work with a driveshaft in place. After years of building cars, he feels that the angle between the pinion and the driveshaft is the most important. "When I can, I just put a driveshaft in place between the engine and the rearend and then move the rearend housing until I have the pinion at the same angle as the driveshaft - and then I point the pinion down another one or two degrees."

When there is no driveshaft to mount between the transmission and rearend, Jim sets the pinion angle at the same angle as the engine - minus one or two degrees (nose pointed down). The idea is to make the U-joints "work" as they rotate. If everything were in perfect alignment the cross shaft would always rest against the same needle bearings in the cap. When the angles are different by a few degrees the cross shaft will walk around the caps, rotating the needles and spreading the load and wear as it does so. A difference of a degree or two is also a good way to dampen vibration and harmonics that can build in the driveline and resonate through the car.

Serious horsepower requires a heavy duty U-joint. Whether your shaft is aluminum or steel buy one with yokes for big U-joints.

Measuring the angles

Measuring the angles of the shafts can best be done with a simple protractor. While some transmission and rearend cases have areas where a good measurement can be made, the most accurate readings are made across the yoke itself or across the caps of the U-joints.

The final position should leave the engine positioned far enough back to allow for good cooling system components and good air flow. It should be positioned so the belt-driven fan will pull air through the center of the radiator. The rearend should mount so the pinion angle is slightly different than the angle of the engine centerline.

The rear end should have at least three inches of travel from ride height to the compression stops, and there should be one inch of clearance between the tires and wheels, and the suspension components or inner fender lip.

The well known and often photographed Boyd rear suspension is based on the earlier, pre-'84, Corvette center section with a billet rear cover and hub carrier. In-board brakes make maintenance more difficult but reduce un-sprung weight.

Chapter Eight

The New Body

Fiberglass or Steel

BUYING A BODY

This chapter is intended to provide an overview of reproduction bodies, both steel and fiberglass. A chance for you to consider what's out there, how they are built and what differentiates a great body from an OK body. Also included are tips on mounting the body and making it ready for paint.To make the chapter as comprehensive, and fair, as possible we've included Q&As with owners of both a 'glass and a steel body manufacturing facility.

One of the most popular of the reproduction bodies available in both steel and fiberglass is of course the '32 Ford. Henry must be smiling because there are probably more Deuces on the road now than in late 1932. Harwood

Whether you choose an original body or one of the reproductions will depend on your taste in cars, your budget and your own skill level. A skilled body man might think nothing of repairing a beat-up-but-solid steel body, while the rest of use shudder to consider what the labor bill would be to straighten and repair the old steel hulk. Original tin has that certain allure, but you have to wonder if it's worth it to spend all the money and time necessary to restore an old original body.

Fiberglass molds must be continually updated and repaired to ensure a quality body. Harwood

Even though two Deuce roadster bodies from two fiberglass manufacturers might look the same at first glance, there usually are significant differences in the two products. Difference in the details, differences in the way they are reinforced, even some subtle differences in the manufacturing process. Which is not to say one is good and the other bad. It's more a matter of finding the best body with the features you want at a price that makes sense.

An employee of a large street rod manufacturing company suggested that in fiberglass, "you get what you pay for. There's a reason that the best bodies cost two thousand dollars more than the inexpensive ones. You can judge the fiberglass body by what it fits. It should fit the original frame, not a specific frame that the body manufacturer or someone else manufactures."

In buying reproduction bodies you tend to get what you pay for. Shop around and buy a quality body, and be absolutely sure it will match up to your intended frame.

Fiberglass manufacturing

A new body built from fiberglass starts with a mold. Generally these molds are formed around a good body or "plug." Obviously the quality of the reproduction body is greatly affected by the quality of the original. There is some loss of detail in the manufacture of a

New from Harwood is this three window '32 Ford with or without running boards. Harwood.

'glass body. Like a second-generation photograph some slight loss of detail is inevitable.

Molds are generally formed of tool-quality fiberglass that is laid up around the plug and then reinforced with a framework of wood or steel. Manufacturing a body or fender begins when the resin and catalyst are sprayed or brushed into the mold. Fiberglass cloth can be laid up in the mold by hand or sprayed in with a chop gun (a gun that sprays a slurry of chopped fibers and resin).

Building a fiberglass body in our industry is generally a job done by hand. Finished quality is very dependent on the skill and care exhibited by the workers, the quality of the body used to make the mold and the quality of the mold itself. Differences in quality between one manufacturer and another become more important as you look at more and more fiberglass products. As you look at the various bodies and components at The Nationals, keep in mind the following points.

1. How thick, and strong, is the fiberglass.

2. How flat are the panels.

3. What kind of framework is used inside the body.

4. What about the fit and finish? How smooth is the gelcoat? How do the various components, doors and trunk lids, fit the body.

If you want equal time for Chevy products, consider the range of Chevy cabriolet bodies available from Downs Manufacturing, including this 1935 roadster. Downs

Q&A: Gary Harwood

Q: Gary, why don't you start by giving us a little background on yourself and Harwood Industries.

A: I started Harwood 24 years ago. I like to say that in 24 years I've gone from one employee (me) to over 80, working at two facilities here in Tyler, Texas. We are probably better known to the race car crowd, we make race car bodies, fuel cells, Lexan windshields, even hood scoops. In the last four or five years, however, we've gotten real serious about street rods. We just finished another new building at our second facility and that will allow us to add substantially to the number of street rod bodies we can manufacture without affecting our other business.

Currently we make the '32 Ford roadster, in either a hiboy or full fendered configuration. We also have a three window '32 coupe. Next will come a '32 sedan. We also build our own frame for the '32. We have rails manufactured to our specifications and then we build from there. We intend to stay in the '32 line. Our goal is to be the best at what we do. We don't try to have something for everybody.

Q: What about fiberglass and steel? From your perspective, what are the advantages of buying fiberglass.

A: Fiberglass is so much easier to modify. If you want to fill a seam or close up a hole you don't have to have any welding skill. It's just easier. And some of the steel bodies need a lot of preparation before you can put paint on them. I think the fiberglass and steel bodies are essentially going to different markets. Some guys are just "steel guys" while others just want a certain body and they want a good product.

Q: There seem to be more and more fiberglass manufacturers out there. If I'm shopping for a popular body, how do I ensure the body I buy is a good one? What do I look for as I compare bodies?

A: Look for fit and finish, and overall quality. If you look at the car hard enough you can see how it was manufactured. You have to look at a lot of different bodies. Some are pretty crude in the details because the tooling they use is crude. Check where the inner and outer door glue together. Do the two panels match up? How are the doors hung, do they fit? Look to see if the manufacturer has had to make repairs to the body before it went on display. Is there any primer or body filler in the body? Check the structural integrity of all the bodies.

Our customers are informed. They typically are not first-time buyers. The people that buy our products have had several cars, sometimes several '32s. When I talk to people, I tell them to educate themselves. Look over as many bodies as possible, the raw cars. And talk to owners of the various bodies. Would they buy from the same company again? Did they have any trouble? Was the company there to help?

Q: Some bodies are reinforced with wood while others are reinforced with a steel framework. Are they equally good?

A: Wood is not a reinforcement, period. Commonly people put wood in there to provide a tack strip for the upholstery, but it's not reinforcement. Wood only works to strengthen a body if it's entrapped in the fiberglass - not just glued in place.

Steel reinforced bodies are also laughable. In an accident that light tubing doesn't do any good. Our bodies do not have steel or wood to reinforce the body. We use a honeycomb composite material and combine that with good design. All of our seams are radiused, for example. We use compound radiuses in all the corners to eliminate stress. The strength should be in the structure not in the framework.

Q: How much hand work goes into the manufacture of a good fiberglass body?

A: We make bodies by hand-laminating the fiberglass, we don't use any chopper guns. Any good body is made by hand lamination. And our crews are very expe-

Another reproduction Deuce, this one in steel from Brookville Roadster Company. Brookville

rienced. We also let the bodies cure for two to three weeks.

Q: What kind of preparation is required before these bodies can be painted, and are there any tricks to painting fiberglass?

A: The first step is to wipe the body down with a good wax and grease remover to remove any wax left from the molding process or any finger prints. Then block sand with 80 grit and spray with a high-build epoxy primer-surfacer. Then block sand again as necessary. Continue blocking and spraying, progressing to 180 and finer sand paper as you would for any body. The primer-sealer you use should be compatible with the finish-paint system you intend to use. There are no special tricks to painting our fiberglass bodies.

Q: Do all bodies fit all frames? And are there any tricks to mounting the body on the frame?

A: Our bodies will fit an original frame or a good reproduction frame - any good body should. A few of the bodies just aren't right. They may be knock offs of knock offs and the dimensions have changed.

When you mount our body on a frame, it can require a few shims between the body and the frame. We like to see people start at the front, at the firewall edge and get that level so the hood gaps are even on either side. Once the cowl is flat on the car we move toward the back. If there's a need for shims or to move the body a little bit on one side we like to see that at the back corner of the car because it doesn't affect anything else. We do recommend that people use a body welting between the body and frame as it helps to keep the moisture out. Nylock nuts are a good idea when assembling fiberglass but of course our frame has the nuts built into the frame so you can assemble it without a helper on the other side.

Q: What are the mistakes people make when they buy a fiberglass body?

A: They don't do enough research. Our roadster body is sixty nine hundred dollars, they're out there for a lot less. We like to see customers get educated. They need to know that at the end of the project they've got a lot of money in this car so they shouldn't skimp on the front end. Two or three thousand dollars isn't very much difference when you consider the cost of the total car. I tell them, 'save up, buy the best body you can. Do as much research as possible. That way you're a happy hobbyist five years down the road.'

Q&A: Kenny Gollahon @ Brookville

Q: How about a little history on the Brookville Roadster Company?

A: My dad was a sheet metal apprentice and an avid hot rodder. He started out making patch panels for some of his cars. Pretty soon all of his friends wanted him to make panels for their cars and then one year he took a pickup truck load of parts to Hershey, Pennsylvania and sold out there. He started the original company, Antique Auto Sheet Metal in 1971. The parts were all Model A patch panels and we just kept adding panels to our line until we pretty much had a whole body, minus the cowl and a few other pieces. So then we started making the complete Model A roadster body. The '32 body was a little different because

Among the more popular products from Gibbon is their '39 Ford convertible body, available already installed on a Fatman chassis if you so desire. Gibbon

there probably aren't enough Deuces out there to justify making patch panels. With the Deuce we started by making the entire body right away.

Q: Can you walk us through a typical manufacturing sequence for a complete body?

A: All the parts start as a flat sheet of cold-rolled steel. The gauge varies from 16 to 20, depending on the parts we're making. The gauge always matches exactly the gauge that Henry used. First the parts are laser cut into blanks, then the blanks go to stamping. If there's any trimming to be done we have a six-axis robot that does the trimming.

After trimming it goes to secondary tooling. This is where we might fold the edge of a door or a quarter panel for example. Next we put the bracing inside the body and then pre-assemble the panels. We build the complete body on a fixture that mimics a factory frame. We put in the sub-rails and then work up to the cowl and the rest of the panels.

Q: How accurate are your bodies compared to the originals?

A: The bodies are very accurate. The gauge is the same, we guarantee that all our parts will fit an original body and that our bodies fit an original frame. It's interesting that when we were tooling up for the Deuce, we had 8 original doors and no two of them were exactly the same. The dimensions varied by up to a quarter inch. You have to realize that the original parts were made in a number of different facilities and that the standards of today are much tougher than the standards of 1932.

Q: What do I look for if I'm in the market for a steel body?

A: Look at the fit and finish, the quality. Our bodies pretty much sell themselves.

Q: Are there any tricks to mounting the body on the frame?

A: Not all the aftermarket frames are the same, that's why we guarantee our bodies will fit factory frames, or our own. We like to see welting used between the body and the frame. Otherwise it's pretty much a matter of trial and error, experience helps a bit too. Some shimming will be required, just like putting any body on a frame.

Q: How much preparation work is required before a Brookville body is ready for paint.

A: Most guys will spend a day or two on the body prepping it for paint, though of course some will spend more because they want this perfect show car.

Q: What kinds of mistakes do people make when they buy a body, any body?

A: They take short cuts, in both buying and finishing. You might save a thousand at the start of the deal but it will get you in the end. Sometimes a guy will buy our Model A body and then put it on the original frame. Which is fine but sometimes by the time they disassemble the frame, clean it up, install an X-member and get it all back together, they could have just purchased a good frame from us or someone else for less money and a lot less time.

A look at the busy shop at Harwood Industries. In addition to bodies they also manufacture their own '32 chassis. Harwood

Chapter Nine

Wiring and Accessories

Amps, Volts, Ohms & Other Mysteries

The Basics

We offer here a brief overview of hot rod electronics, including an overview of the main components in any electrical system and some options when it comes time to buy a harness or harness kit. Much of this information is taken from the Wolfgang book, *Hot Rod Wiring.*

Before describing the circuits in your hot rod, we first need a few terms and their definitions.

Voltage: the force that pushes electrons through a wire (sometimes called the electromotive force).

Current: the volume of electrons moving through

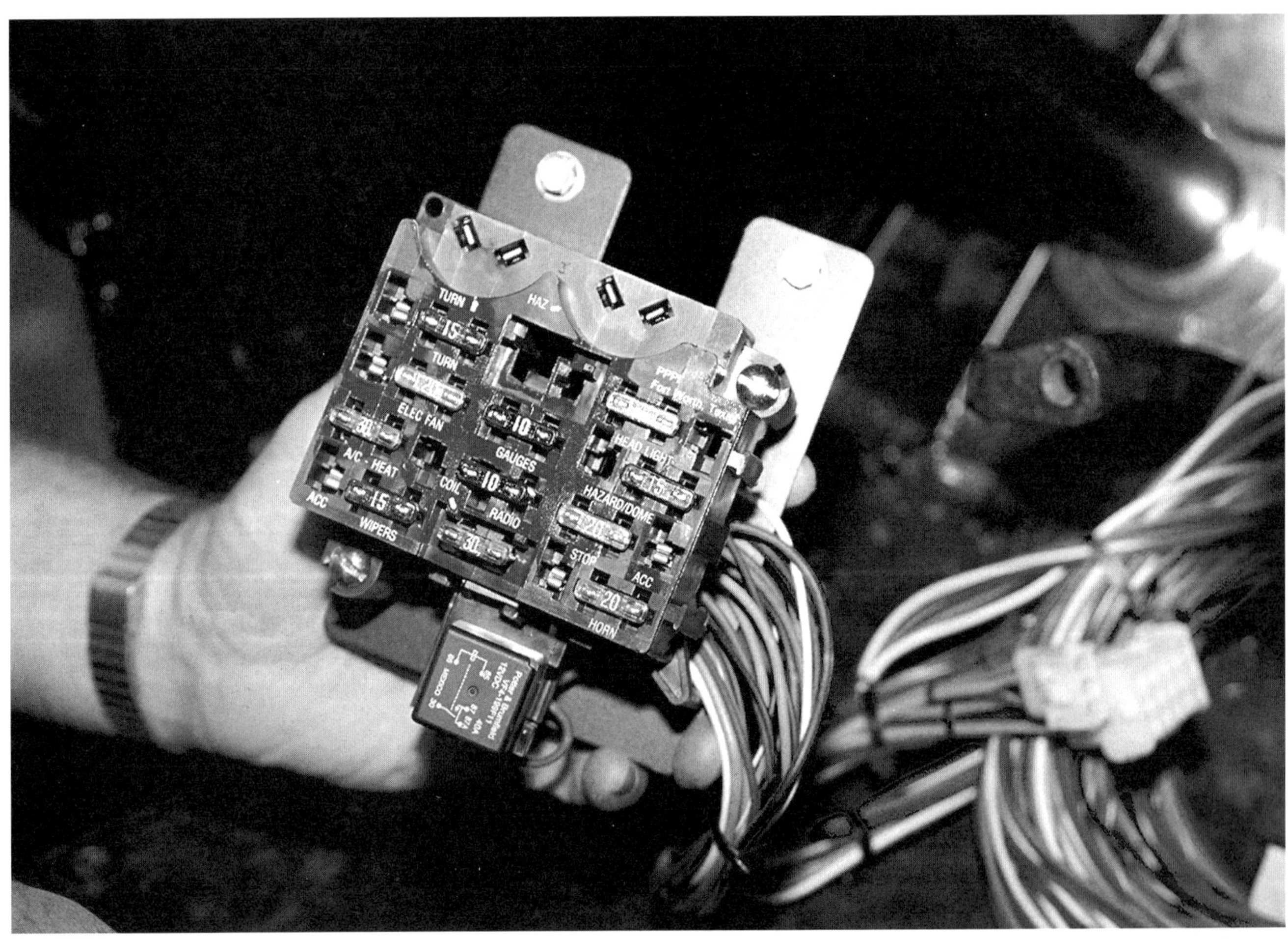

What used to be the hardest and most intimidating part of building a car - the wiring - has been made much easier thanks to the wide range of wiring harness kits which have come on the market.

the wire measured in amps.

Resistance: the restriction to the flow of electrons measured in ohms.

The way these three forces interact is contained in a simple formula known as Ohm's law.

Ohm's law: V=IxR or stated another way, I =V/R and R=V/I What this law really means is that voltage, amperage and resistance are all related. You can't change one without changing another.

Length	0-4ft.	4-7ft.	7-10ft.	10-13ft.	13-16ft.	16-19ft.
Current						
0-20A	14ga.	12ga.	12ga.	10ga.	10ga.	8ga.
20-35A	12ga.	10ga.	8ga.	8ga.	6ga.	6ga.
35-50A	10ga.	8ga.	8ga.	6ga.	6ga.	4ga.
50-65A	8ga.	8ga.	6ga.	4ga.	4ga.	4ga.
65-85A	6ga.	6ga.	4ga.	4ga.	2ga.	2ga.
85-105A	6ga.	6ga.	4ga.	2ga.	2ga.	2ga.
105-125A	4ga.	4ga.	4ga.	2ga.	2ga.	0ga.
125-150A	2ga.	2ga.	2ga.	2ga.	0ga.	0ga.

Wire size chart from the IASCA handbook. The size of wire used in any circuit is determined by the current load and the length of the wire. This chart is for copper, don't use aluminum wire.

The importance of wire size

The size of wire is very important in the flow of electrons. The larger the wire, with more strands in the wire, the more current it can carry. Different wire sizes and types are manufactured with different amounts of strands. Most household wire is made of a single heavy strand, good for carrying high voltages and low current. In the automotive applications the wire is sized from light to heavy and is always made up of many strands, which is good for carrying higher current flows at relatively low voltage. Multiple strands also makes the wire flexible and less prone to breakage from vibration.

The size of a wire is known as its gauge. Bigger numbers indicate a smaller wire able to carry a smaller current. A 22 gauge wire might be used for a dash circuit while a 2 or 4 gauge wire would make a good battery cable.

As a general rule of thumb, always use the highest number of strands per wire size (or gauge) as possible. A good example of this is battery cable. Many people use welding cable for their battery cables because it has much higher stranding than regular battery cable. This results in less voltage drop and heat build up during use, due to the lower resistance of the multiple strands.

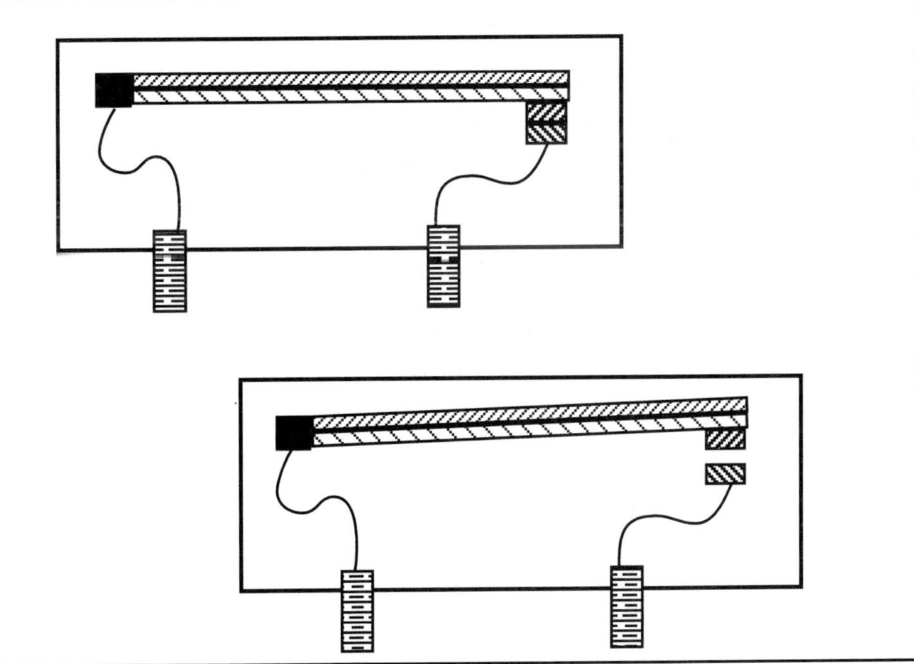

Too much current moving through the bi-metallic strip on this circuit breaker creates heat which causes the dis-similar metals to expand at different rates and lift the upper contact off the lower contact.

Wire size in a circuit

Charts are available that give the recommended gauge size for a particular situation (see the wire size chart). The two things that determine the gauge needed for a circuit are the current load the wire will need to carry, and the length of the wire that carries that load. More current requires a larger diameter wire (smaller gauge number). The same current, but in a longer piece of wire, will require a larger diameter wire. When in doubt, go larger in size, not smaller.

Some of the best wire is rated TXL with insulation that is thinner, yet more heat and abrasion resistant, than anything else on the market. The thinner insulation means it can be hard to gauge the gauge of the wire. What looks at first like a 16 gauge wire might actually be 14 gauge wire with the new, thinner insulation.

Circuit protection devices

Fuses

A fuse is one of the most important parts of the electrical circuit. The fuse is the weak link in the passage of current and is designed to allow only a preset amount of current to flow through the circuit. By using a fuse, regulation of current flow is possible and damage to sensitive electronic parts and powered circuits can be avoided.

A fuse works by having a small conductive strip between the two contacts that is designed to melt at a certain temperature. When current flow reaches a certain maximum level, the natural resistance of the strip creates enough heat to melt the strip, thus stopping current flow. If a wire rubs through the insulation and contacts the frame, the fuse will blow well before the wire gets hot enough to melt. Without a fuse you run the risk of melting the wires in one or more circuits and starting a fire in the car.

Fusible links

A fusible link as used by the factory is a short link of melt-able wire housed in high temperature insulation. Generally used to carry a load heavier than a standard fuse can handle, fusible links are again the weak link in a chain, designed to melt before the wire or circuit itself are damaged. Detroit often uses a fusible link where a large feed wire (eight or ten gauge) connects to the starter solenoid or source of battery power. If the wire with the fusible link in it shorts to ground someplace "down stream" the link will melt preventing the entire wire from burning up.

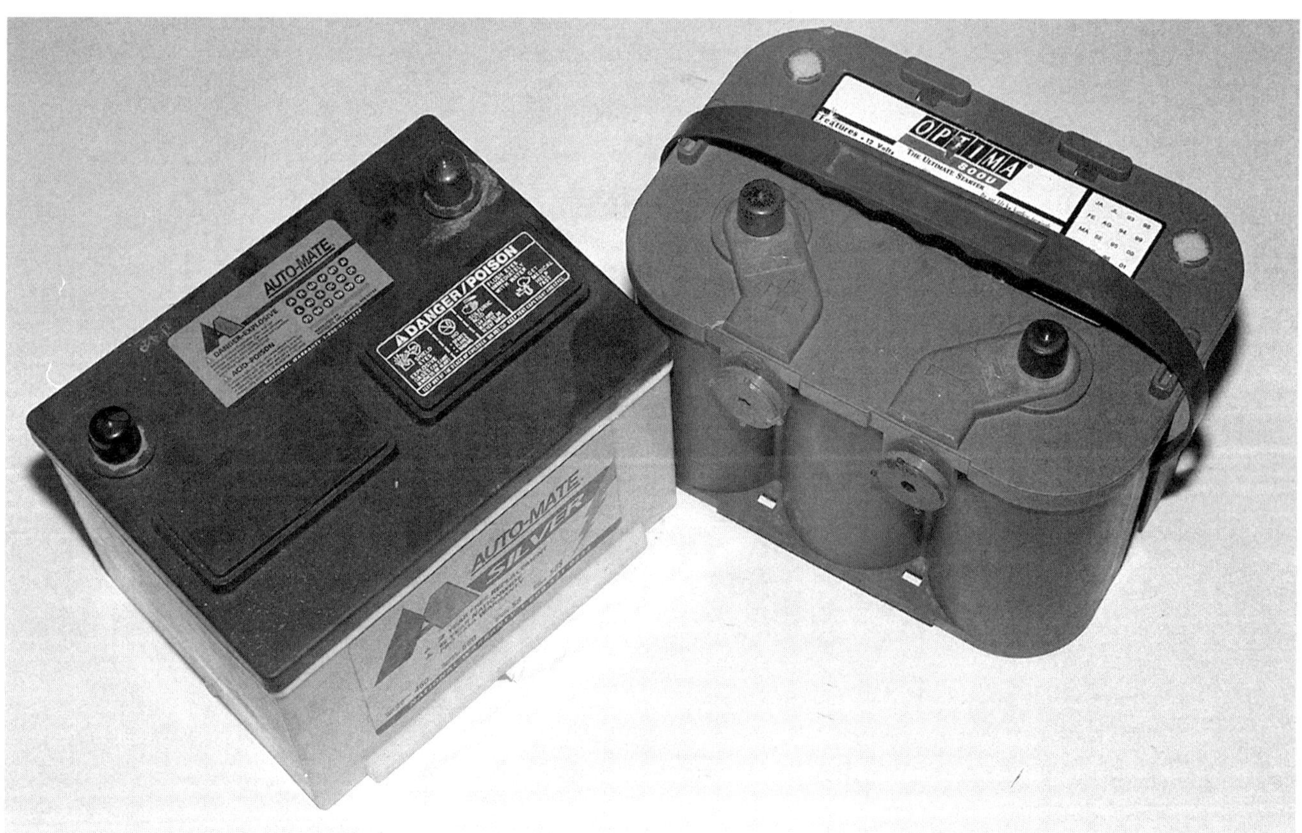

Batteries come in standard 24 Group sizes and the newer, fully sealed Optima recombinant designs.

Circuit Breakers

Circuit breakers, like fuses, are designed to protect circuits from overloading. The major difference between fuses and circuit breakers is that fuses, when overloaded, must be replaced. Circuit breakers, by contrast, can be re-used after the short circuit or overload is repaired. Most of the circuit breakers used in automotive applications are spring loaded so they automatically reset after cooling off.

BATTERIES, STARTERS, ALTERNATORS.

The battery might be called the heart of your electrical system and serves three major functions.

1. The battery is an electricity producing device. The chemical reaction between the lead plates and the electrolyte, a water and sulfuric acid mix, creates electrical current. The voltage is determined by the number of cells.

2. The battery is also a storage device. The battery can store a large amount of current in its plates and is capable of providing this current to the electrical system on demand.

3. The battery is also a regulator of current in the system. As the engine rpm or system loads increase or decrease, the voltage and current flow go up and down. The battery acts as a buffer to damp out spikes and stabilize voltage in the system.

There are two basic configurations for automotive batteries: top post and side post. The side-post battery was developed to help solve several problems. Because the posts are not next to the caps or vents, a source of acid fumes, corrosion of the terminals is reduced. The added benefit of the side-post terminal location is the lowered silhouette which allows the batteries to fit more easily into modern cars with lowered hood lines.

Batteries carry a number of ratings, the two most common include cold cranking amps and reserve capacity. Cold cranking amps is the amount of current the battery can provide for a certain length of time at a given temperature. To determine the rating a battery is chilled to zero degrees Fahrenheit and placed under a load in amps for 30 seconds while maintaining a voltage of 7.2 volts. The larger the rating number, such as 500 / 600 / 750, the more power that battery can put out to start your vehicle.

The reserve capacity rating is a means of determining how long a battery might supply current in a situa-

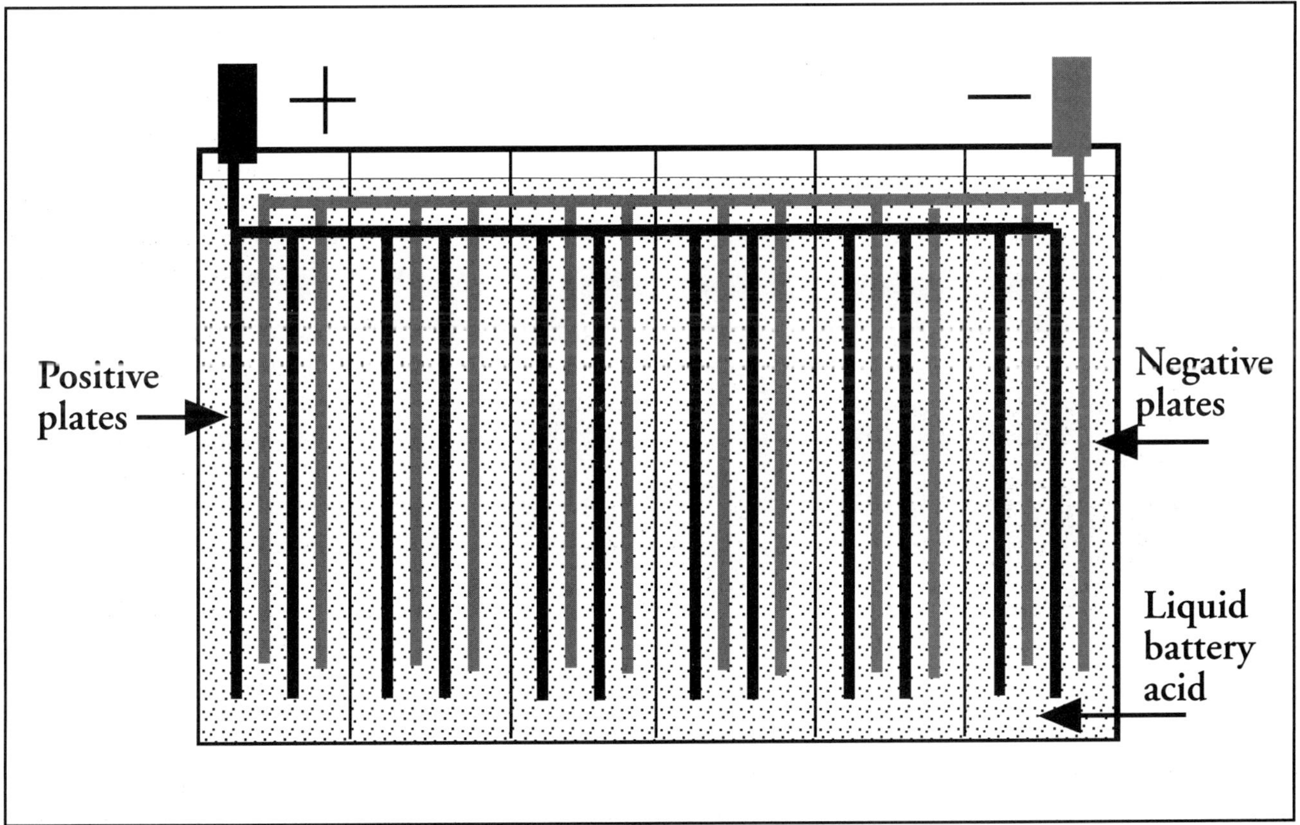

The typical battery is made up of positive and negative plates suspended in a acid bath. The new Optima batteries wrap the positive and negative plates into spiral cells.

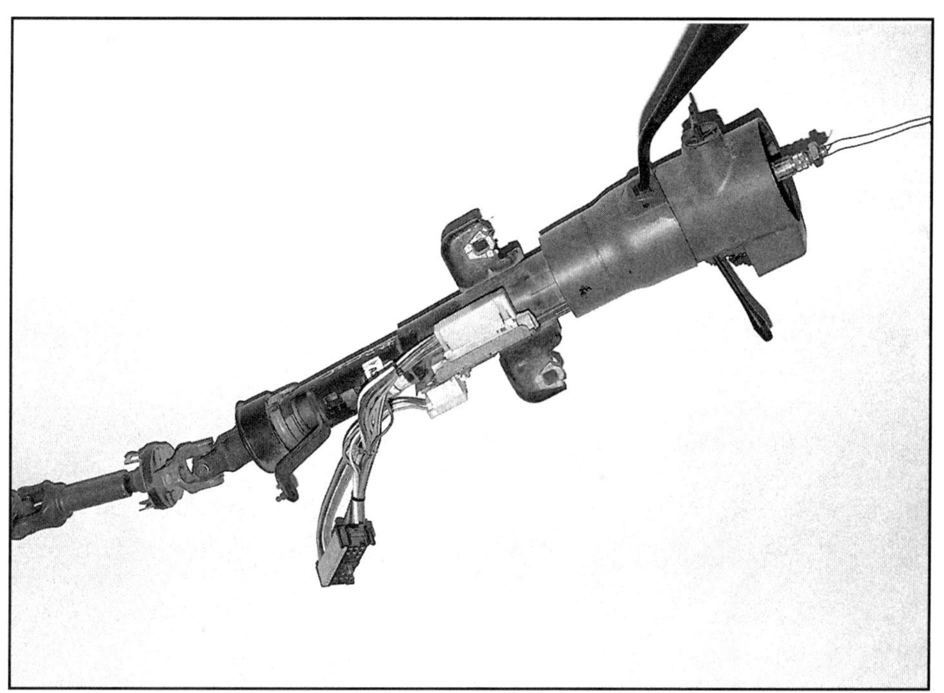

When you buy a steering column, be sure to get the ignition and dimmer switch, and both the male and female ends of the plug-ins.

tion where the charging circuit has failed. This is the length of time, in minutes, that a fully charged battery can be discharged at 25 amps without allowing the individual cell voltage to drop below 1.75 volts.

Basics

Automotive batteries are constructed of positive and negative plates kept apart by separators, grouped into cells, connected by straps and suspended in a solution of electrolyte. Each cell of the battery produces approximately two volts, by connecting six cells in series a 12 volt battery is created.

Nearly all batteries emit hydrogen gas, explosive to say the least. Gassing is especially likely when the battery is being charged by an external charger, but also when the battery is under a load. For these reasons cigarettes, sparks and flames must be kept away from the battery. When jump starting a car be sure the last connection to be made is the negative cable, connected to the frame or engine block of the car being jumped. That way any spark occurs away from the battery.

Wrenches laid on the battery have the potential to short across the terminals and create an explosion. Also remember that metal jewelry conducts electricity at least as well as a wrench (silver and gold are both excellent conductors), which is one more reason to take off the watch and rings before you start work on the hot rod. Remember too that batteries contain sulfuric acid, corrosive to metal and damaging to human skin. Spilled acid should be flushed thoroughly with water.

The alternator you buy must be matched to your car, in terms of output, mounting brackets and also the style of belt pulley.

Battery Chemistry

The plates of the battery are made of lead alloys. Specifically the positive plates are made of lead peroxide while the negative plates are made of sponge lead. These plates are suspended in a solution made up of sulfuric acid and water. When the battery discharges, sulfate (sulfur and oxygen) from the electrolyte combines with the lead on both the positive and negative plates.

As these sulfur compounds are bound to the lead plates, oxygen is released from the positive plates. The oxygen mixes with hydrogen in the electrolyte to form water. As this reaction continues the acid becomes weaker and weaker, and more and more sulfate coats the plates. Charging the battery reverses the chemical process, forcing sulfates back into solution with the electrolyte and causing oxygen from the solution to move back onto the positive plate.

Don't buy your alternator on looks alone. Be sure the output is sufficient to run all your accessories and then some.

The down side to all this charging and discharging business is the inevitable flaking of lead particles off the plates until the battery's ability to act as a battery is greatly diminished. Further affecting battery performance is the fact that when a battery is left discharged the sulfates penetrate too deeply into the lead plates and cannot be driven back into solution, creating the condition often referred to as "sulfated."

Specific gravity is often used to check the state of charge for non-sealed batteries. Specific gravity simply measures the weight of a liquid as compared to water. The specific gravity of a fully charged battery (with strongly acidic electrolyte) ranges from 1.260 to 1.280 at 80 degrees Fahrenheit, or 1.260 to 1.280 times as heavy as the same volume of water. As the battery becomes discharged the specific gravity drops because the electrolyte has a higher and higher percentage of water. This is also why a discharged battery will freeze on a cold winter's night

The big output wire on the alternator must be large enough to handle the full output of the alternator. Some builders put a fusible link in this wire because it runs direct to battery power.

while a fully charged battery will not.

Low and no-maintenance batteries change the chemical and physical construction of the plates slightly. By adding a chemical like calcium to the plates and changing the structure of the plates themselves, gassing of the battery is greatly reduced. This means a much smaller volume of corrosive/explosive gasses, little or no loss of water, and generally improved performance.

Recombination batteries, sometimes known under the name, valve-regulated, go even further. These batteries contain all the "electrolyte" in a porous glass mat positioned between the cells. There is no liquid acid in batteries of this type and they can be mounted in nearly any position. As a General Motors engineer explained, "the chemistry is the same as a conventional lead-acid battery, but the hydrogen and oxygen can move back and forth from the plates to the glass mats without the gassing you see in a standard "flooded" battery design.

Sometimes confused with recombination batteries, the new gel-cell batteries are quite different internally. Though they are often used for golf carts or stationary applications these designs are currently not suitable for automotive applications because they are so easily damaged during recharging.

Deep cycle batteries

Golf Cart batteries are known as deep-cycle, meaning they are designed to be discharged to less than half their capacity with no ill effects. A standard automotive battery by contrast is designed to put out a short burst of high-amperage power, to start the car, and then be quickly recharged by the alternator. Using an automotive battery in a deep cycle application will shorten its life considerably.

Keep it charged for long life

All batteries self discharge to some extent. This means a fully charged battery will draw itself to zero voltage over time, even if the battery cables are removed. The answer is to recharge the battery when the vehicle sits for any extended period. This becomes doubly important with the new fuel injection and radio designs which place a small load (but still a load) on the battery at all times. Don't allow the battery to run down and don't let it sit for any length of time in a discharged condition.

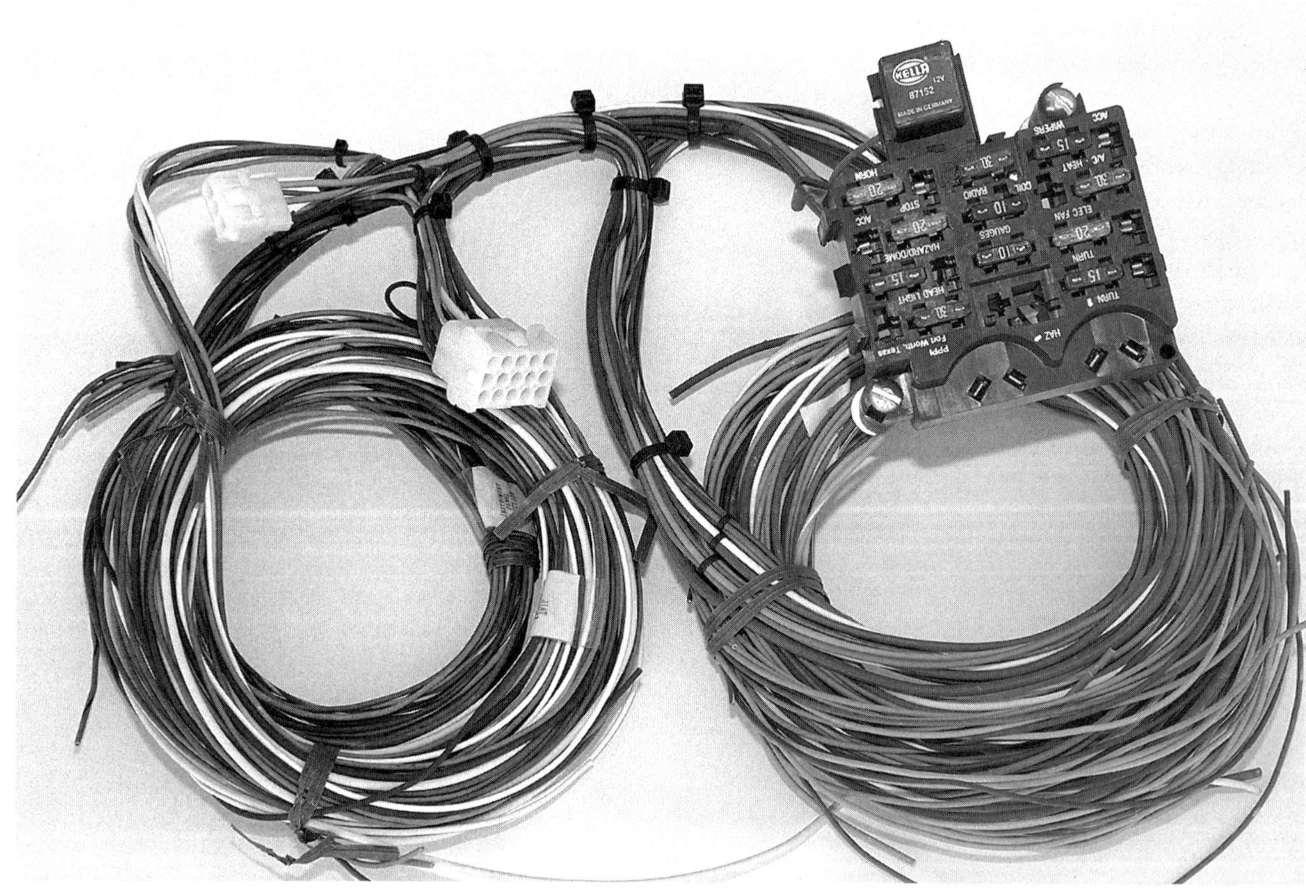

Many wiring harness kits come in various levels of capacity. A simple hot rod only needs a 12 circuit harness while a fat-fendered cruiser with A/C and a killer stereo probably needs closer to 18.

Other tips for long life are mostly common sense. Keep the battery clean because a film of acid and dirt will conduct a small current between the terminals, speeding the self discharge.

Mounting the battery

Batteries, with the exception of the truly sealed recombinant designs, emit gasses that are very corrosive and explosive. When mounting any battery be sure to keep it away from high heat areas like exhaust pipes and manifolds. It's also necessary to keep the battery away from any source of sparks, such as the ignition system, that could accidentally ignite the gasses vented by most battery designs.

If you decide to build your own harness, fuse boxes like these are available though it's a lot more work than buying a complete kit.

The closer the battery is to the starter the better it will perform in starting the engine. Different vehicles have larger or smaller areas to mount batteries in and often the battery for a hot rod is mounted in the trunk. In these situations it's good to remember that the farther the battery is from the starter, the larger the battery cables need to be in order to prevent excessive voltage drop. Though it's common practice to use the frame as the main conductor on the ground side of the battery-starter circuit, it's a better idea to run a ground cable from the battery negative post to the engine.

The use of a marine grade battery box to house the battery will protect the trunk from acid spills and also protect the battery case from damage due to items carried in the trunk. Some of the standard batteries use a vent tube, (Delco for example) which can be run through the floor if the battery is mounted in the trunk or interior of the car. Otherwise, the battery box itself must be vented to the outside or the battery must be one of the

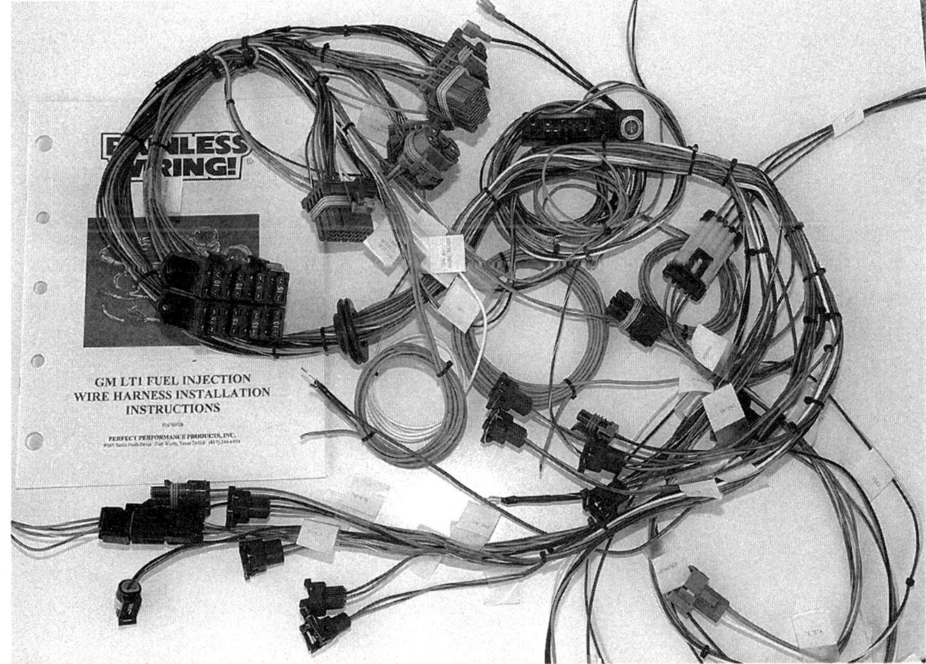

You can convert many fuel injected Detroit engines to hot rod use with the help of harnesses like this one designed to interface with the factory computer, injectors and hardware. Painless Wiring

Two very commonly used starters, a typical GM with attached solenoid on top, and a Ford starter on the bottom.

truly sealed recombinant designs.

Battery cables are heavy so it is important to route the cables in a way that provides them with plenty of support. Running them inside of the frame or solidly attached to the outside of the frame will prevent them from drooping and rubbing on a sharp frame edge or a bracket. The last thing you want is to have the cables wear through their insulation, or come in contact with the drive shaft or the exhaust system. Use plenty of brackets and ties to keep the cable(s) out of harm's way.

Starters

Factory starters can be broken down into three basic styles: Those that use a solenoid attached to the starter like most GM designs; the bare starter with the solenoid mounted to the fender or firewall, often used by Ford; and the gear reduction starter used by some Chrysler products, which uses both a solenoid attached to the starter and a starter relay.

The typical Ford starter circuit uses the separate solenoid seen on the left while GM starters use the solenoid on the right. In either case, the solenoid is really just another relay.

At the heart of all starters is a powerful electric motor. Most starter motors are made up of two or four field windings inside a heavy steel case. Rotating inside the field windings is the armature, wound with many windings of copper wire. Each armature winding is attached at the copper commutator, where the brushes connect the various windings to the battery and ground as the armature and commutator rotate.

When the starter is energized, current moving through the field windings creates a strong magnetic field with a north and south pole. Current moving through the armature creates a magnetic field in one or two of the windings. This magnetic field in the armature winding is attracted to or repelled by the magnetic poles of the field windings. It is this attraction and repulsion of magnetic forces, between the field and armature windings, that causes the armature to turn. As the starter turns, different commutator strips line up with the brushes and a new set of windings are energized, putting continual torque on the rotating armature.

Despite the tremendous torque created by most starters, enough to crank over a big block V-8 on a cold morning, some engines require even more. Even without the need for more torque, some applications call for a smaller starter that provides more clearance between the starter and some nearby component like the headers or steering box. In answer to these needs we now have the high torque mini starter.

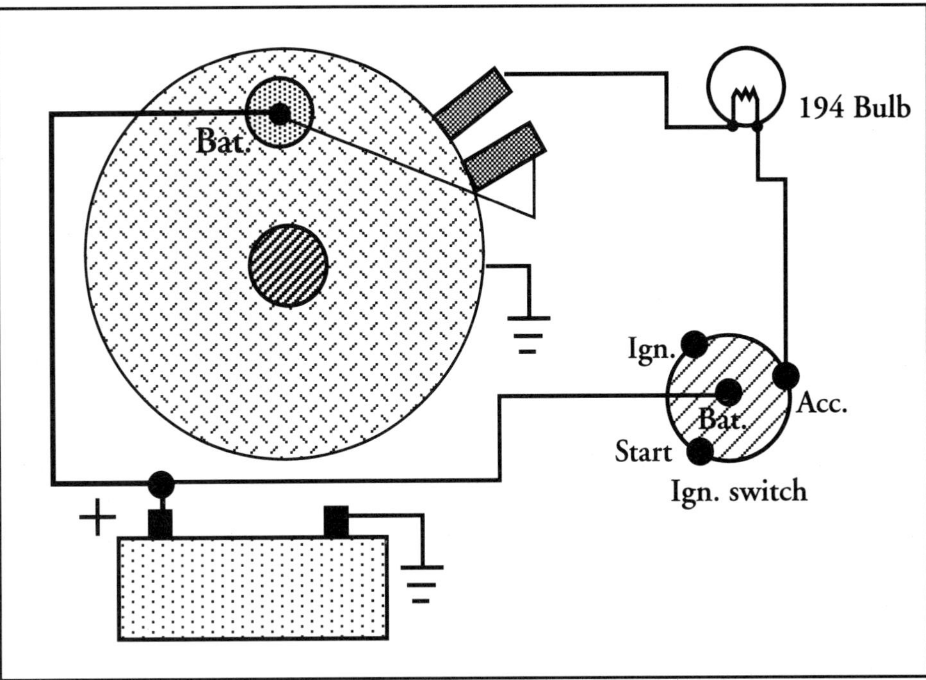

This diagram shows the easiest way to wire an early GM alternator. By wiring the exciter wire off the accessory position (instead of ignition) you eliminate run-on when the key is shut off - caused by voltage fed back from the alternator.

The high torque mini starter is being used in all types of applications. First designed for racing applications, the mini starter is popular for the hot rod builder because of its small size and extra torque. The size means more clearance around the starter and better air circulation so it is less likely to be affected by heat from the exhaust headers or manifold. The extra torque helps to turn over engines with extra displacement or compression. Many of these designs use gear reduction to help obtain the extra torque.

When installing a starter, remember that the battery cable needs to be large enough to carry the current and routed so as to avoid the manifolds and edges of the frame. The connection between the battery cable and the starter or solenoid must be kept clean and tight. As already mentioned, the starter

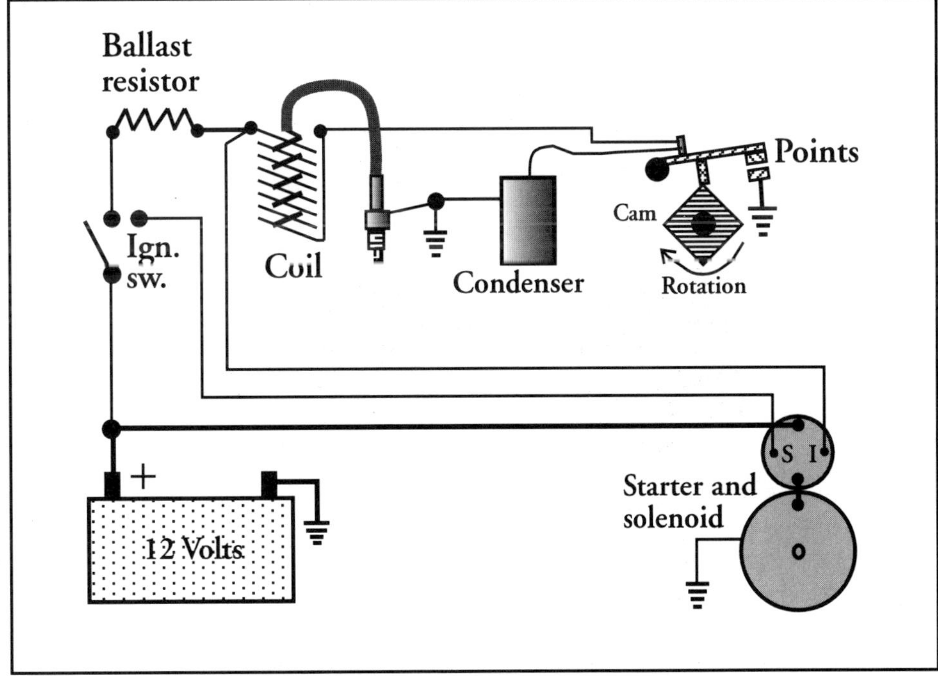

This circuit shows a typical points-style ignition circuit, with the bypass circuit used for starting. Modern electronic ignitions replace the points and condenser with a magnetic timing device.

must have a good ground. In most situations that means the cable from the negative battery post should go directly to the engine block.

Starter solenoids

Starter solenoids are essentially the switch between the starter and the battery.

Two main types of solenoids are used: the direct mount where the solenoid mounts directly on the starter (GM style), and the remote mount, where the solenoid is mounted on the fender well or frame and a cable connects it to the starter (Ford style).

The direct mount solenoids are more likely to encounter hot-start problems due to heat build-up from the exhaust system. Solenoid failure, or non operation, is often due to excessive heat. As heat builds up so does resistance. The extra voltage and current required by the solenoid to overcome the resistance is, at times, not available because the battery cables are too small or the connections are dirty and corroded.

A number of devices have been offered over the years to help owners overcome this hot-start problem. The most common and useful is the kit made up primarily of a standard 30-amp relay that is located in the starter switch circuit. When the ignition switch is turned to start the relay is activated and in turn transmits power from the battery terminal of the starter directly to the S terminal on the solenoid. This is, in effect, a safe way of shorting the two terminals together with a screwdriver.

Charging systems

The battery provides energy to start the car. Once it does start that energy must of course be replaced. The charging system is designed to bring the battery back to a state of full charge and provide the power needed to run all the electrical systems on the vehicle.

Two different types of charging systems have been used in automobiles. Up until about 1960 most cars used a generator, while from that point on most cars relied on an alternator to keep the battery charged and provide energy for the vehicle electrical system.

A generator produces direct current voltage (DC) and uses a regulator to control the current and voltage output, and to connect the generator to the battery.

Alternators produce alternating current, (AC), in which electrons flow first in one direction and then in the other. Since DC is used in automobiles, the AC current coming out of a raw alternator must be converted to DC. Diodes, often mounted in the alternator frame, convert the AC to DC electricity. Unlike generators, alternator current output is self regulating. Voltage, however, must be regulated. Though the first alternators used an external regulator for voltage, most alternators used on late model automobiles use an internal voltage regulator.

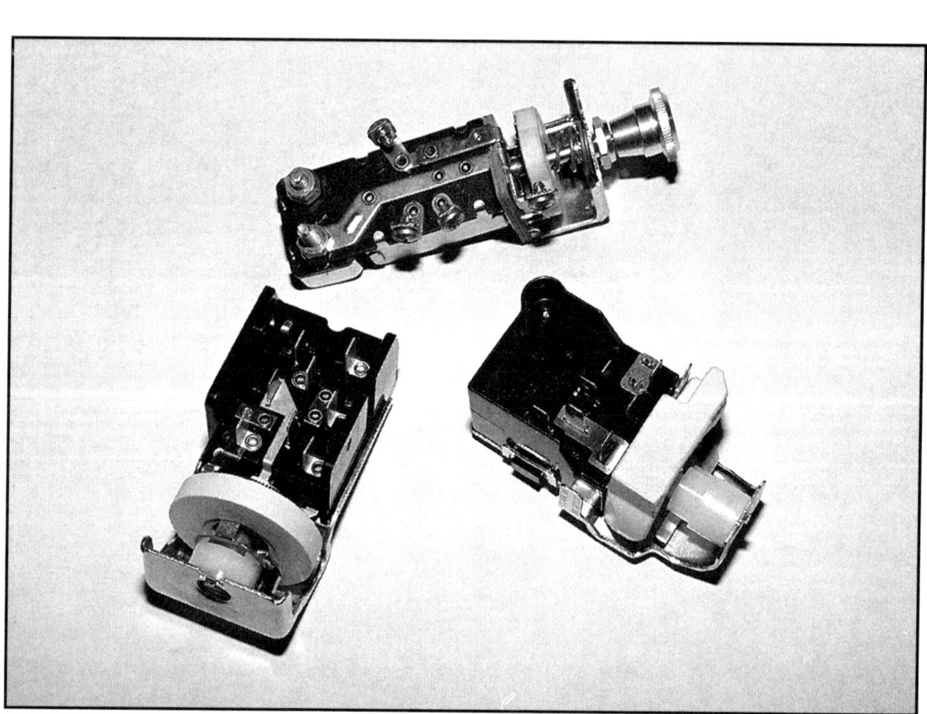

Switches for headlights can be either O.E.M. style or a universal as seen on top. Most universal switches include a circuit breaker, seen on the end of the switch.

Note: All alternator systems sense system voltage in order to determine the correct voltage output of the alternator. Thus a bad battery will make it impossible for the alternator to work correctly and may also make testing of the alternator difficult. Because it's so important that the regulator sense the true condition of the charging system, it's very important that the alternator or generator and any external regulator be well grounded. If in doubt use a star washer under the regulator base or run a separate ground wire.

Construction Details: Alternators

Alternators contain the same basic components seen inside a generator, though the components are essentially reversed in position when compared to a generator. The field windings are actually contained in the rotor, the component that spins; while the stator (which takes the place of the armature in a generator) is stationary and mounted to the alternator case. Because alternators produce alternating current (AC), two banks of diodes are used to convert this to direct current (DC).

Detroit switched from generators to alternators for good reason. Most alternators produce more power than a generator, and are able to provide a high percentage of that power at a relatively low speed. In addition, alternators will more readily spin to high rpm without damage. And though both generators and alternators contain brushes, the small alternator brushes run on the smooth slip rings of the rotor and thus last for a very long time.

Because factory alternators don't have a permanent magnet as part of their field windings, they will not self-excite. That is, no matter how fast you spin an alternator it will not produce power until the field is energized by an outside voltage source.

By adding permanent magnets, however, to a GM alternator some aftermarket companies have created the "one wire" alternator. Because these units are internally regulated they need only one wire, the output wire, which typically runs to the battery. The "one wire" alternator has become very popular with hot rodders due to the ease of installation. The main disadvantage to these alternators is that a dash warning light can not be wired in, and the use of a voltmeter or ammeter will be required. Also, most of these are based on earlier GM alternator designs so total output is not as high as the more modern alternators.

Note: the large wire on the back of any alternator is hot all the time, thus many factory installations include a insulating cap of some type where the wire attaches to the alternator. Also note the output wire needs to be sized large enough to carry the full output of the alternator. In most cases this means *at least* a 10 gauge wire.

How to choose an alternator

One question that often comes up when building a car is: "How big an alternator do I need?"

In answer, remember that bigger is not always better. An alternator rated at 100 amps output will seldom put out more current than one rated at 65 amps. The output ratings are in direct relationship to the engine rpm. The larger, in size and output, alternators normally only have maximum outputs at their maximum rated rpm range.

When choosing an alternator, determine first the amount of current needed by the system during full load conditions. The size of the alternator you choose should be close to that amount. Remember that it is very rare that all the circuits in the vehicle will be turned on at one time.

Plenty of hot rods are equipped with General Motors alternators simply because the drive train is likewise based on GM components. Among the newer GM alternators is the CS model, used on many GM products starting in about 1987.

Smaller than the earlier alternators, these units are readily available in the junk yard or at swap meets and come in outputs as high as 140 amps and sometimes more.

This alternator offers a number of advantages for a prospective street rodder looking to replace an earlier style GM alternator or to simply install a late-model alternator.

The shaft size is the same as earlier alternators,

Relay kits are available for accessories like cooling fans that draw too much current to be run straight off a switch.

which means you can install either a V-belt or serpentine pulley. The mounting lugs are very similar to earlier alternators, which makes upgrades relatively easy. And wiring is a fairly straight-forward proposition.

Charging system wiring

The wires for the charging system are as important as any wires in the vehicle. The feed wire from the output post of the alternator to the battery must be capable of carrying the maximum output of the alternator, which may be 80 or 100 amps. Often this wire is too small and won't handle the heavy current load. If the insulation on this wire melts off and the wire touches ground you're likely to have a complete electrical melt down. If you check the IASCA charts, a 50 amp circuit that's four feet long requires a 10 gauge wire, and 50 amps is a minimal alternator output these days, many go to 100 amps and more.

Alternator exciter wires are usually small in gauge size because they are only transmitting a signal to the regulator, telling it when to charge and how much. They should, however, be at least 16 gauge to prevent too much voltage drop.

Wiring harness kits.

Harness kits built for hot rods and competition vehicles come in every creed and color. The simplest consist of nothing more than a fuse block and a wiring schematic. The best have the wires already attached to the fuse block, feature labeled wires and have enough circuits to handle air conditioning, electric fuel pump, power windows and a heavy-duty stereo.

Though there's always a temptation to "do it ourselves," it's not always a good idea to buy the ultra basic kit and then try to wire the vehicle ourselves. Unless you're a fully trained automotive mechanic with electrical experience, the wiring of even a simple street rod is pretty complicated. What's more, a Mickey Mouse wiring job is one of the things that's sure to give you trouble when you roll that new ride out of town. And while it might be easy to fix a leaky carburetor in the middle of Iowa on a Sunday afternoon, finding and repairing an open circuit or a short circuit while parked along the super slab is no fun at all. This last scenario also tends to make you very unpopular with anyone else in your little caravan.

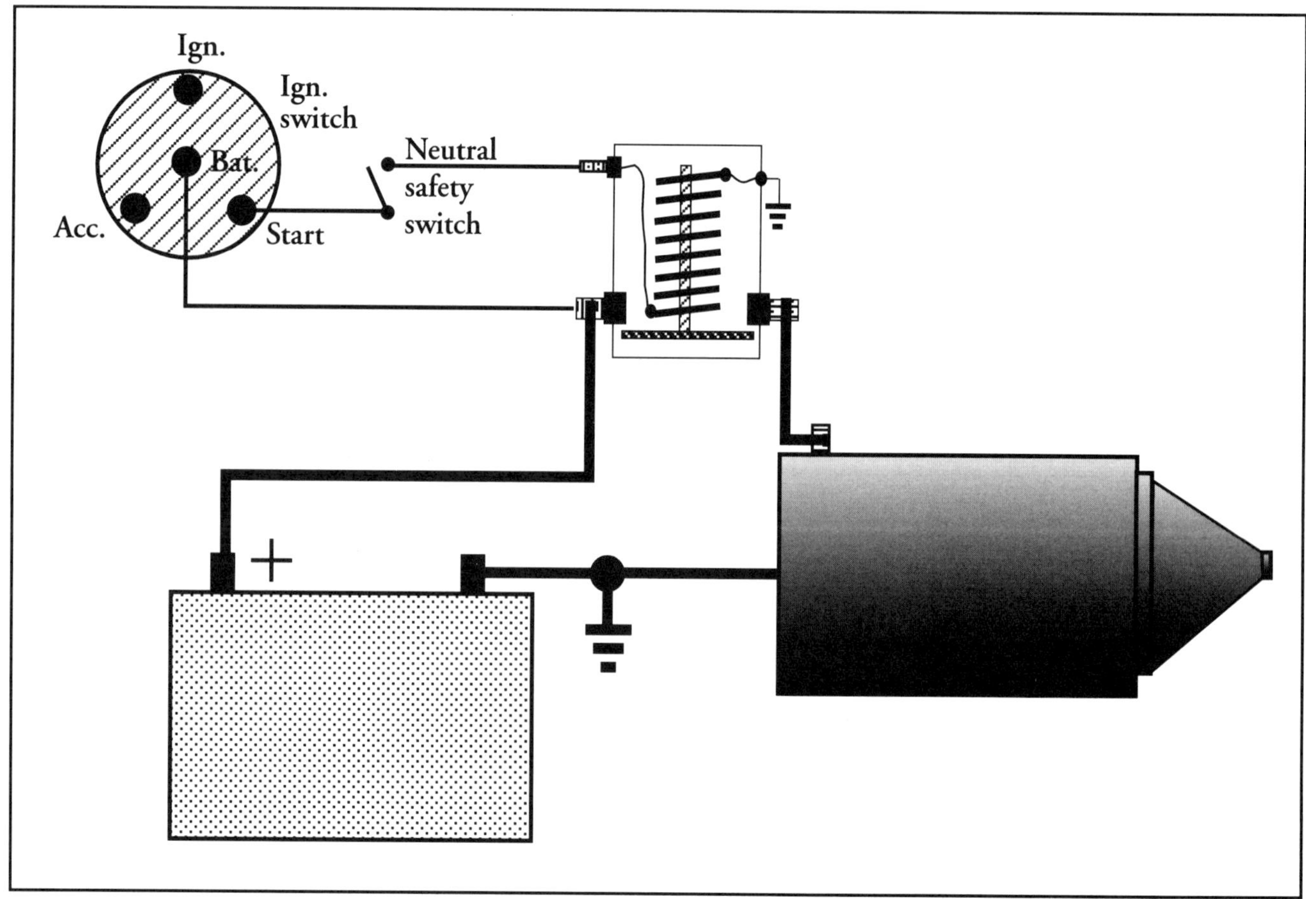

A typical Ford starter circuit. When you hit the key the disc in the solenoid is drawn up, which acts as the switch between the battery and the starter.

What to buy

As always the first thing to ask yourself before buying a wiring harness or harness kit is, "how will I use the car, and how many circuits and accessories will this car have?" For competition vehicles there are specialized and very simple harness kits that provide extra circuits for electric cooling fans and fuel pumps and no circuits for air conditioning.

If your street rod is a barebones roadster, then simplified harness kits are available with only enough circuits to handle the basic duties of starting, ignition, lights and a few accessories. Forty Ford sedan owners with all the electrical goodies can step up to a kit with 18 or more circuits, so there are plenty of circuits and no need to add an auxiliary fuse panel or use one circuit to power two accessories.

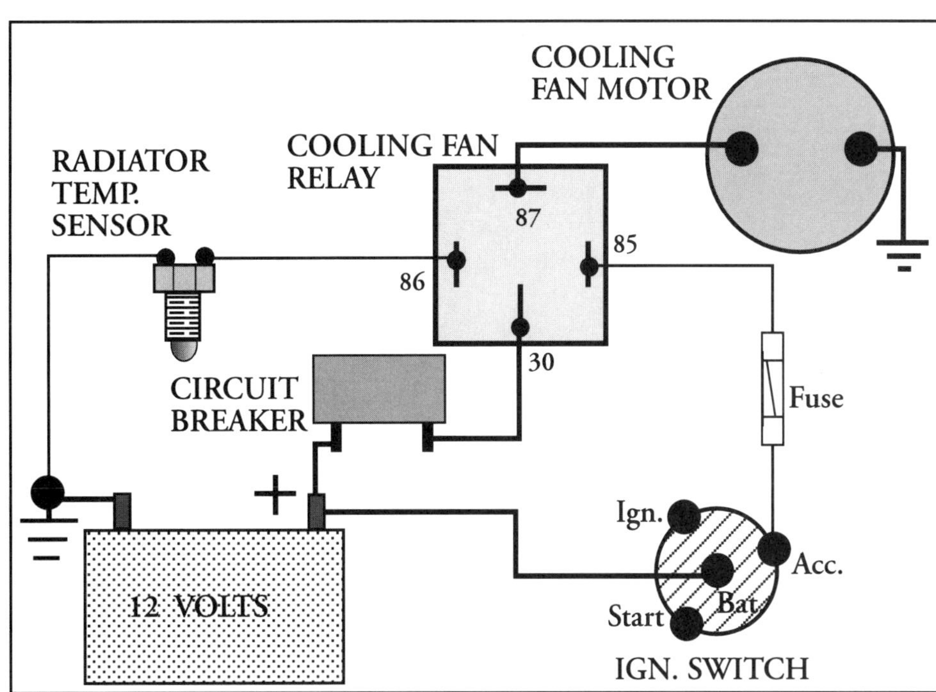

In this cooling fan circuit the sensor is used to trigger the control side of the relay, while the actual current that runs the fan moves though the load side of the relay.

Consider the overall quality of the harness kit, the length of time the company has been in business, and the availability of a 800 help line. Before buying, call the manufacturer and explain what it is you're building. If the new ride is a sedan with a GM steering column and a drivetrain from a late model Firebird, ask how their kit will interface with those particular parts. Many of the technicians on the other end of the 800 numbers are very experienced and can offer suggestions as to which column or ignition switch to use with which harness kit.

A number of companies make OEM-style harness kits for later model hot rods like '55 Chevys and early Camaros. For most street rod applications however, a somewhat more "universal" harness kit will have to be used.

Electrical wiring kits vary by design, some require more connecting than others. At least one firm ships the fuse block with all the wires attached to the block. You need only run the wires to the proper components and attach them there. Other systems provide a fuse block and separate labeled wires. These require that you run the wires from the electrical components back into the car and then attach them to the fuse block. Look around and call around before making your purchase.

No matter which system you buy into, take time to study the instructions and the kit before jumping into

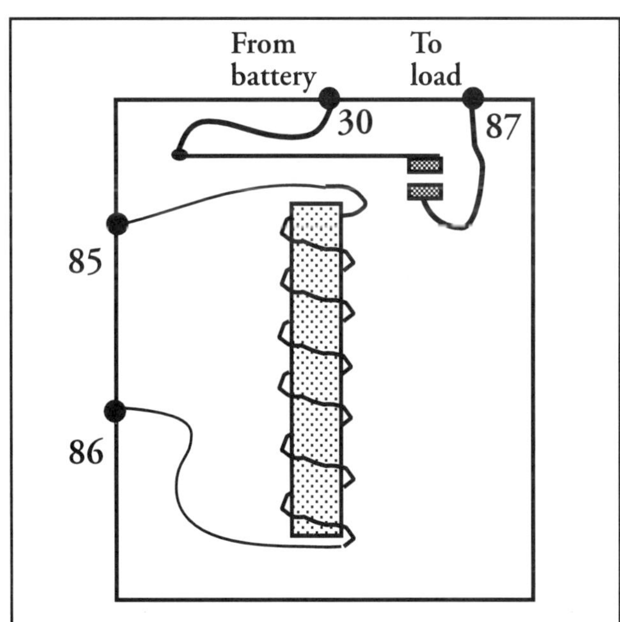

A typical relay uses the control side (#85 and 86) to create a magnetic field, which then pulls down the armature and closes the load side. Solid-state relays work the same though the mechanical parts have been eliminated.

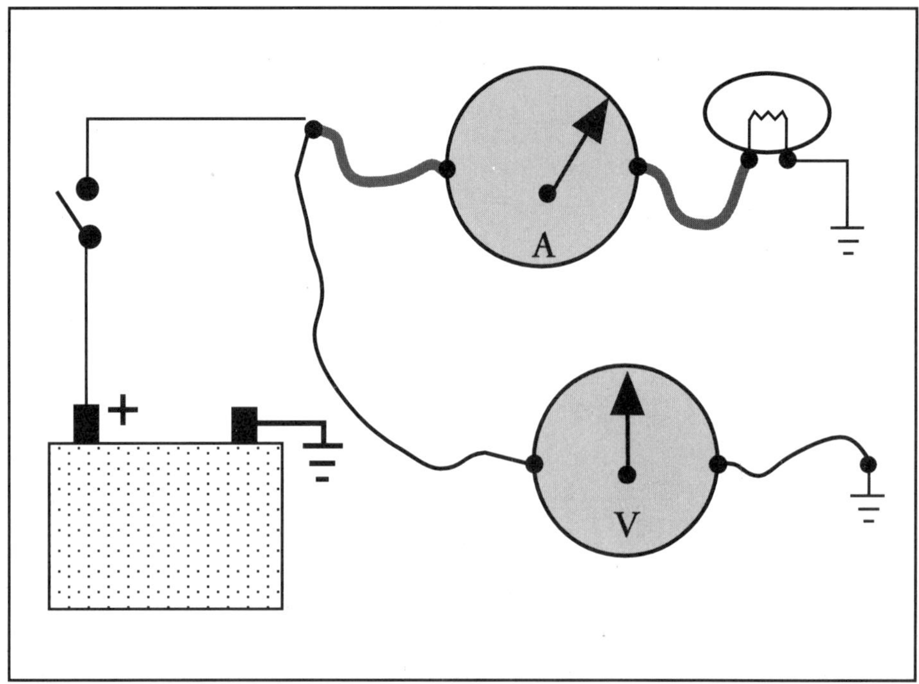

Ammeters and voltmeters are very different critters. Ammeters must be wired in series while a voltmeter is wired in parallel.

the project. Wiring comes near the end of the project and you may be anxious to finish the car. Don't let that anxiety force you to rush through the job. The product engineer from one of the harness manufacturing companies suggested that the best thing to do with the new wiring kit was to take it into the house, brew a pot of coffee and spend an hour or two studying the kit and the instructions.

Perhaps the best part of the *Hot Rod Wiring* book is the trouble-shooting section, a chance to talk with a technician who gets people out of trouble all day long. The following is a question and answer session with Dave McNurlen, the man who answers the tech line at Painless Wiring. Before coming to work at Painless, Dave spent 20 years as a GM technician and is rated as an ASE certified Master Technician.

Before we could get to the most popular questions - and their answers, Dave offered the following two tips for anyone installing a harness kit in any vehicle.

First, Dave explains that many of the calls to the tech line could be eliminated if people would just read and follow the instructions. "They should take the instructions into the house and read them from cover to cover - before they start to do the wiring job. It will make the job easier and save time in the long run. The other good idea is to buy or borrow a factory wiring diagram. This doesn't apply to street rods or race cars, but for most of the others it's very helpful if you have some idea what the original harness did. You will probably end up using parts of the original harness in conjunction with the new harness kit, so you need to be able to correctly identify the wires and components."

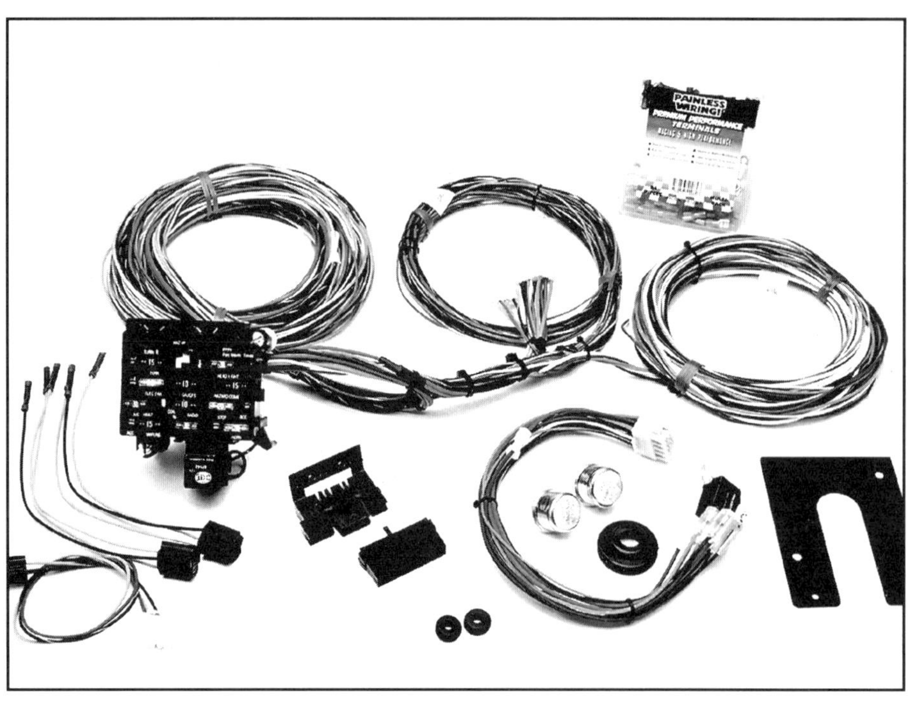

The better quality wiring harness kits include nearly everything you need, including the wires, the fuse block, pigtails for the ignition and light switches, and the bracket for the fuse block. Painless Wiring

Commonly asked questions

Q: Should I hook up an amp meter or a voltmeter?

A: You should hook up a voltmeter. It's safer, you don't have those big wires running through the dash. Voltmeters are more modern, and still have good indication of whether or not the charging circuit is working.

Q : When do I need a ballast resistor?

A: You need a ballast resistor with a points-type ignition. With electronic ignition you don't need a ballast resistor (with the exception of some Chrysler ignition circuits). Some younger people don't understand the old points type system and why it needs a ballast resistor. They don't understand the bypass circuit either.

Q: When do I use a relay, what does it do and why can't I just use a switch?

A: Things like cooling fans that draw a lot of amps are hard on switches. Most switches won't handle that much amperage without premature failure. I explain it this way: the relay is the switch for the load, and the switch itself just activates the relay,

Q: Which alternator should I buy and how do I hook it up?

A: You should figure out what the power demands are that you will place on the alternator. There are a lot of options here. Match the alternator to the loads you're going to put on it. There are a number of different circuits, it's confusing and you need to know which alternator you've got and then follow the wiring diagram for that particular alternator.

Q: Should I buy a one-wire alternator?

A: I don't recommend them. They seem to need more rpm to energize, to start charging, but you don't know that unless you're watching your voltmeter all the time. The newer CS series alternators from GM work well and make good power at very low rpm, The one-wire alternators look pretty, but they're based on the old GM alternator so the output is limited compared to the newer designs.

Q: Where's the wire for the brake lights?

A: For many American cars the brake light switch feeds power to the turn signal switch, the brake lights are actually fed through the turn signal switch - the brake and turn signal light or filament are the same thing. (Note, some later cars with separate red and amber brake and taillights actually run the brake light wire back to the brake light bulbs on either side.)

Q. I've got two cooling fans and the AC compressor on one circuit and I keep blowing fuses. Why can't I just put in a bigger fuse?

A. The circuit will only handle so much current. If you put in a bigger fuse than the circuit is designed for you risk melting the wires in that circuit.

The electric cooling fans should be run off a relay. Mount the relay under the hood. Take the power off the starter, for example, fuse it with a fusible link, and then run the power to the load side of the relay, through the relay to the fan and ground the other side of the fan. I tell people to keep all the heavy circuits under the hood so they have shorter load wires, less voltage drop, and less chance for wires to get chafed or cut.

The farther the battery is from the starter, the more important the battery cables become. Use the very highest quality cables and be sure to solder on any ends. Painless Wiring

Chapter Ten

Tim Allen Roadster

Hand Built in the Extreme

Presented here is a very lovely and unusual car, built by Steve Moal of Oakland, California for Tim Allen of Home Improvement fame. Though it's hard to accurately describe the building of a hand-built car in only sixteen pages, I've tried to do just that. As noted in the Acknowledgements, many of the photos here are the work of Michael Dobrin. The writing is my own, though in writing it I've tried to let Steve tell much of the story himself. Tim Remus

The more complex the project, the more important the sketches and renderings become. This rendering by Don Varner was one of several done for Tim Allen. Steve Moal describes these as "food for thought." A way to determine what your customer does and doesn't like.

The Roadster

The Tim Allen roadster, hand built by Steve Moal, really started with an earlier and in some ways similar, car. Steve built the *California V-8 Special* when he learned of his nomination as Builder of the Year at Oakland Roadster Show. In his attempt to build something that was "different from just building another really nice Deuce hiboy" Steve Moal gave all of us a look at the true depth of his talent as both a designer and a metal-smith.

Early shot of the chassis shows the extremely strong "rails" made of 1-3/4inch chrome moly tubing. Belly pan will wrap under the bottom of the lowest tube. Floor will be positioned on top of the bottom tube. The area between will be foam-filled.

Steve and his stunning black Special received a tremendous amount of attention, especially when the car appeared on the cover of Street Rodder magazine. One of those who called in praise of the car was a certain TV personality by the name of Tim Allen. A long-time motorhead, Tim didn't just call to express his approval of the car. Essentially, Tim Allen called to say, "The car's great and I want one."

Today Steve describes Tim Allen as a "true enthusiast," and "very easy to work with." Going from that first phone call, however, to the nearly finished roadster required a tremendous amount of work on the part of Steve Moal and his crew. With a project like this it's easy to talk about the big jobs, fabricating the frame or the body panels. But building the Roadster from scratch also meant completing about a million small and seemingly insignificant tasks as well.

The radius rods are hand fabricated from chrome moly tubing. Brakes are Wilwood four-piston calipers.

Steve didn't just put a new body on an existing frame. Steve set out to build the entire car. The frame, the body and much of the hardware.

The Halibrand rearend moves up and down on four-bar links. Upper shock mount seen here was changed, note the photo at the bottom of this page.

Straight axle used in front is supported by a single-leaf spring. Panhard rod is used, though it's hard to see. Michael Dobrin

This shot shows the final upper shock mount. Carrera coil overs allow damping adjustment from the cockpit. Suspension design allows for 7 inches of travel. Michael Dobrin

The design

Steve explains that when it came time to actually design the car he contacted artist Don Varner and asked him for a series of renderings. "We had to start with the California Special. That was the car Tim reacted to. Don Varner worked up a set of drawings that we presented to Tim Allen to get his reactions. Based on his reactions we worked up more sketches until we came up with the final rendering. We tried to key into the things Tim really liked. The whole process is a way to test design concepts."

Once Tim approved the shape, Steve asked Don for a full-size set of drawings. And based on those drawings Steve began the actual building. It's important to note that the final car did not follow the renderings right down to every last detail. "Tim Allen trusts us," explains Steve. "Things did change as we got into building the car, but that was OK because Tim gave us free reign to express ourselves."

The Tim Allen car differs significantly from it's evolutionary predecessor. While the first Steve Moal roadster rides on a 95inch wheelbase, the second generation car uses a wheelbase twelve inches longer. Externally the new car is more a sculptured piece. Internally, it's a more modern machine with a seriously fast 351 cubic inch Ford V-8 in place of the Ford flathead used in the Special.

The work begins

Once Steve knew the dimensions of the new car he could start at the beginning, with the construction of the frame. Don't look for stamped mild-steel rails adapted from something else. The side members of this roadster frame are made up of two parallel pieces of 1-3/4inch chrome moly with a wall thickness of .120 inches.

"Race car builders like chrome moly because of its strength and good memory," explains Steve, "but in this application I'm using it because of the quality of the raw material and the fact that this is what I'm used to working with. The .120inch wall thickness is probably heavier than we really need, but the weight works to our advantage. If the car is too light you get an imbalance between the heavy rear axle assembly and the body/frame that's trying to hold it in place."

For each side of the frame Steve created two parallel tubes of chrome moly separated by short vertical tubes spaced about every twelve inches. These two ladders run down either side of the car and taper toward each other as they approach the front axle.

Following the drawings, Steve and his crew formed two large vertical rectangles of the same chrome moly tubing which will support the front and the back of the cockpit area.

After being cut and bent on the hydraulic tubing bender, Steve heli-arc welded all the chrome moly tubing into a complete chassis. A perfectly flat surface table provided Steve a perfect work surface and a way to ensure the frame stayed true and square.

It's interesting to note while many street rodders use grinding wheels and Bondo to eliminate the appearance of a weld Steve has a different perspective: "Those welds, those chassis welds, those won't be dressed or molded in any way. They're part of the visual joy of a hand-built automobile. It's the way that this car is put together. If it's a weld, we like to see the weld. We also like to see rivets and bolts. The means of attaching things adds texture to the surface. It's a big part of the car and an important element of the design.

Steve and his crew didn't just built the frame, they built many of the chassis components as well. Instead of the latest trick billet A-frames, the suspension here is simple, straightforward and traditional. Steve explains that the reason for the traditional suspension goes beyond the relative merits of the latest independent suspension. "There's a lot of old

At this point the motor and transmission are positioned, and the body shape is starting to develop. Now it's time to decide how the people fit. Michael Dobrin

Note how low Steve's feet are, well below the top of the top tube, which really puts the driver and passenger down into this car. The position of the pedals, shifter, steering wheel and column, and seats, all have to be determined. Michael Dobrin

At this point Steve is trying to determine the flow of the sheet metal, and ensure there will be enough clearance under the hood for the carburetor and air cleaner. Michael Dobrin

A Ford T-5 transmission provides the connection between engine and rearend. The aluminum bell housing was eventually replaced with a Lakewood (scatter-proof) bell housing. Plywood mock-up represents the firewall. Michael Dobrin

Note the quick change rearend with aluminum center section, the Panhard rod and the aluminum driveshaft. The battery was eventually moved to RR corner to make room for a larger fuel tank. You can also see two of the "flanges" that facilitate rearend removal.

Steve tests the action of the frame work that will become the trunk lid. Note the flanges on the edges of the tubular framework. Michael Dobrin

sprint car here," says Steve. "The styling and feeling come from racing. It's evident in the nose, that's a race car nose, and the sports car flavor in the back. It's more like a race car than anything else. Part of our goal was to create a two-seater Indy roadster from the '50s or early '60s, but not later."

At the front of the car the crew installs an I-beam axle, supported by a single-leaf spring and hairpin radius rods. The beam axle is described as part of the car's hot-rod theme. The radius rods are fabricated in-house from 4130 chrome moly tubing, "partly for reasons of geometry and partly to create the right aesthetics," says Steve.

The rear of the chrome moly frame is taller than the front. It's interesting to note that the frame tubes actually run above and below the rearend housing. Because the frame rails wrap around both the top and bottom of the rearend, the lower frame rail on either side features a re movable section. So if the rearend housing needs to come out for any reason the mechanics don't have to try and drag it out through the side of the frame.

The rearend itself is a Halibrand quick-change assembly, what you might call the quintessential hot rod rearend. In the case of the Tim Allen car, the rearend is supported by more of that early-race-car suspension. Four-bar links reach forward from the rearend housing to plates welded to the frame, while a Panhard rod running from the left frame rail to the housing prevents any lateral movement.

Though the suspension seems at first like pretty standard hot-rod stuff, there's more here than meets the eye. Steve explains the suspension by first pointing out the shortcomings of most typical hot rod suspensions. " Most '32 Fords have two or three inches of travel. With that limited travel you have to over-spring the car or it will bottom out. You end up with a harsh ride. But with enough travel designed in from the start we can have a nice ride and control the ride with the shocks, not the springs."

With a full seven inches of travel in the rear, the Tim Allen car can ride on relatively soft springs. The shocks absorbers are adjustable from inside the car, so the driver can stiffen or soften the ride by adjusting the shocks instead of the springs. "We will control the ride with shock damping, not with springs," explains Steve. "The suspension design isn't constrained by the lack of travel."

The rear coil-overs are manufactured by Carrera and are currently equipped with springs rated at 175 pounds/inch. These premium suspension components with the aluminum bodies allow Steve to change the springs later if the 175 pound estimate proves to be too soft or too stiff.

With the chassis mostly complete Steve can install the engine and five-speed transmission, which forced them to work through another set of compromises. "The engine is set pretty high in the frame. It's part of our plan to provide plenty of travel and to have a belly pan underneath the car." Steve goes on to explain, "We also pushed the engine back in the frame. It helps the weight distribution and it also allows us to move the radiator farther back. Remember that the frame gets wider as it goes from front to back. By moving

In this shot you can see the anit-roll bar that runs though the frame.

All the sheet metal panels are formed on the English wheel and each one requires a series of test fits. When the shape is correct the edges of the sheet are "hemmed" around the flanges at the edge of the framework. Michael Dobrin

The steering arm shown here is an early mock up version. Note how the first nose has evolved to the sculptured shape seen here, designed to "reduce the visual mass."

Here you can see the shape of the cockpit developing. The doors will be formed with additional tubing. Before being skinned, however, they will first be cut loose from the rest of the framework and hinged.

The 351 Ford V-8 is set rather high and as far back as is practical. By moving the engine back Steve shifts the weight to the rear and creates more room for the radiator.

the radiator back we can make it bigger. You must have adequate cooling. If this thing sits and idles on a hot day I want it to run out of gas before it overheats. But by moving the engine back the bell housing starts to pinch the driver's feet, that's the trade-off."

Once the engine location was set, the crew could begin mapping out the exact location of the firewall, cowl, seats and pedals. Plywood bucks are cut and shaped to represent the firewall, dashboard and rear body panels. A pair of simple "seats" are set in the car so Steve can spend time sitting in the cockpit, trying to determine exactly where the pedals should go.

"Pedal position is very important," says Steve. "And you have to be sure there's room for your fanny. Basically, you have to be sure people fit in the cockpit. What's different about our package is the fact that our side panel covering the frame rails is eight inches tall. On most street rods your feet would be on top of that eight inch rail. With Tim's car your feet are at the bottom. You really sit down into this car."

Once the driver's position is determined Steve and his crew can begin the process of forming the body panels. The process is one Steve learned from repairing hand-built Italian cars in the family body shop. "We use a method copied from the Italians" says Steve. "They call it *Superleggera* or super-light. It means the body panels are framed in light weight tubing. You know, people say the Italian cars are rough, but if your get in there and look at the things they did, without any of the tools that we have today, it's impressive. Not only are the cars beautiful, the craftsmanship is lovely as well. It's the kind of work I'd be proud to call my own."

"For Tim's car we used 1/2 inch tubing, attached sheet metal 'flanges' at the edges and then hem the body panels over the flanges. Those half-inch tubes are very important because they determine the actual shape of the body."

All the panels on the Tim Allen car are the result of patient passes through 'the wheel' either by Steve or Jimmy Kilroy, an employee at the Moal body shop who Steve calls, "the best metal man and designer I've ever met."

If you ask Steve about his preference for the English wheel, he's quick to explain that there's more than one way to form a trunk lid. "To me it's not very important which tools a person used to form a panel, people shouldn't get stuck on that. Some use a wheel, some like a big power hammer and some use a mallet and a bag of shot. Anyone who's interested should look at what other guys do as a point of interest, and understand that they don't have to do it in exactly the same way with exactly the same tools. People should use what works best for their situation."

Steve and Jimmy formed the panels from aluminum sheet .050inches thick, identified as 3003, a fairly soft alloy of mostly pure aluminum that's easy to work. Each panel took multiple passes through the wheel, followed by some hand planishing, before it was folded over or "hemmed" to the sheet metal flanges at the edge of each piece of framework. Large panels are made up of smaller component panels, which always left Steve trying to decide where the seams should go.

"We try to map out where the seams should go so the panels are manageable, so one guy can wrestle them around." Steve goes on to explain that all the seams are welded in an old-world tradition. "We gas-weld all our seams. It provides good penetration and you can still work the panel later, it doesn't work-harden the aluminum."

Even though Steve works from a full size set of drawings, the shape of the car does evolve as it goes

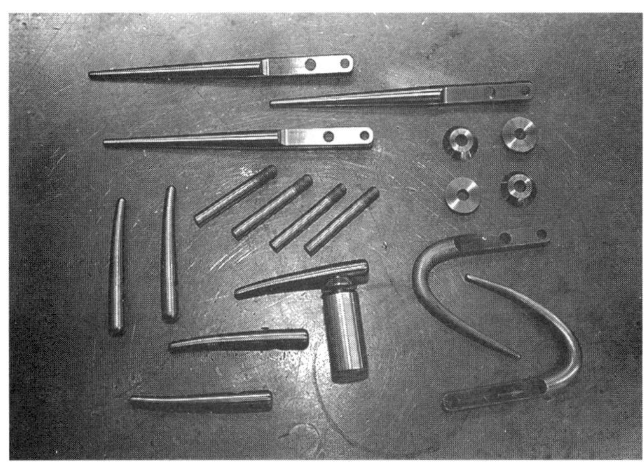

You can tell a great deal about a project by looking at the details. These nicely formed pieces are parts of the over-center trunk latch, with extras. Tim Remus

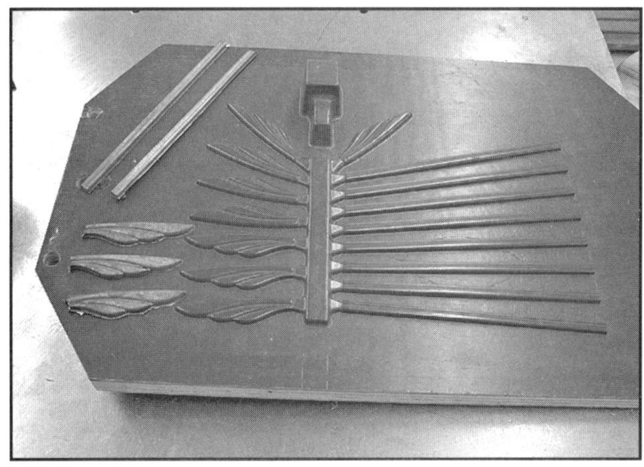

More detail work, the pattern made by Larry Phalen to cast the embellishments used on the dash. Tim Remus

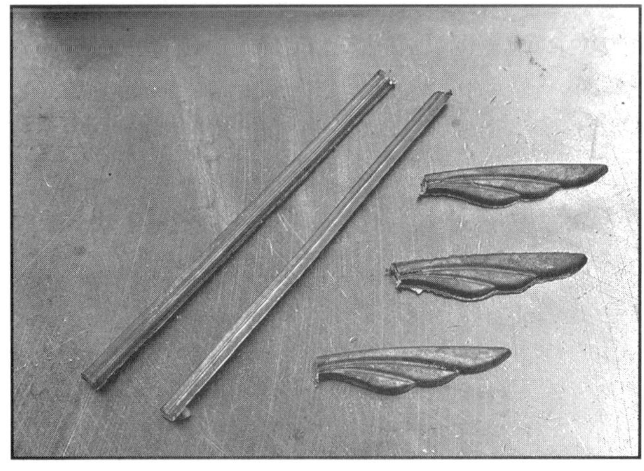

Cast in brass, soon to be chrome plated, these wings will be positioned across the dash of the new car. Tim Remus

The vertical U-shaped tube provides support for both the steering gear and the cowl.

along. It's inevitable that some parts don't look just right once they are on the car and visible in three dimensions.

Steve takes pains to explain that just because you threw a part away and made a replacement, doesn't mean you made a mistake. "The nose went through three transformations. The first one was like the nose on the black car, because that's where we started. But when we got it on Tim's car, which is bigger than the black car, it seemed too big. We decided to sculpt the nose to reduce the visual mass. It took three tries before we got it just right. It's OK to change a part like the nose, it's not OK to not do something about it."

Building a complete car is an enormously complex undertaking and the process doesn't always move in a strictly linear fashion. The various jobs of building the chassis, forming the body panels and developing the steering linkage overlapped. Thus the job of layout and design of the steering linkage took place concurrently with development of the body panels and a hundred other tasks.

In the best race-car tradition, the Tim Allen roadster uses a genuine sprint car steering gear manufactured by Lee Manufacturing. Unlike the Vega gears or rack and pinions seen on most current hot rods, this gear is located up under the cowl, supported by the rectangular framework that also supports the body.

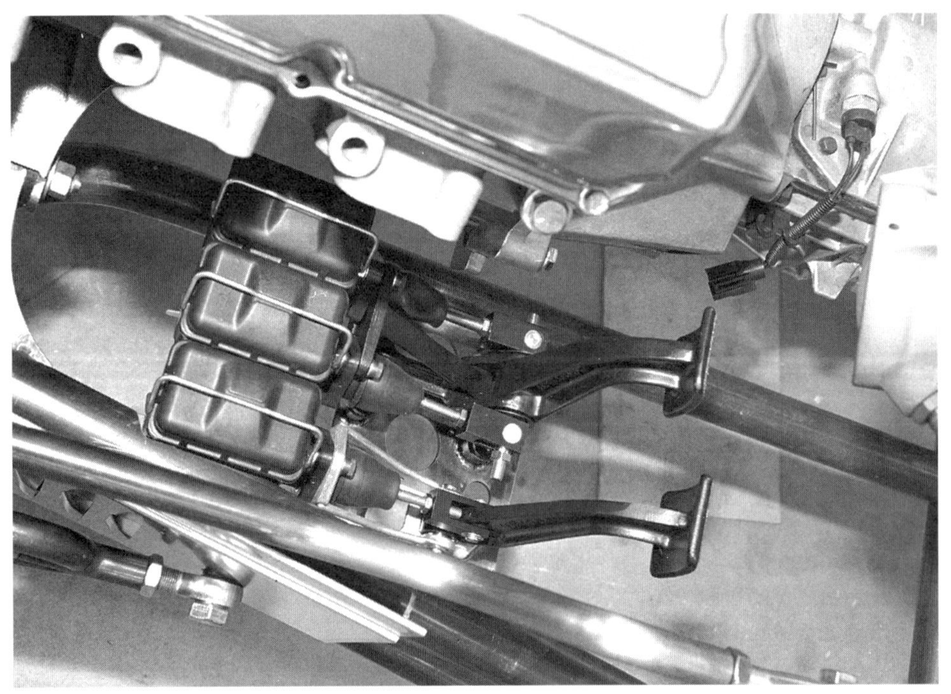

Three master cylinders is a tight fit here between the engine and the frame rails. The outside master is for the clutch while the other two are for the front and rear brakes.

The unusual location of the steering gear works in more than one way to help the design of the car. In Steve's words, "By putting the steering box under the cowl I don't have the box in front of the motor. This means there's more room for the big radiator. And the drag-link style steering linkage is a good system with that great race-car look"

The Lee Manufacturing gear uses power assist. Because these gears are made in small volume for demanding customers, the valving is adjustable and the company is happy to help Steve provide the perfect balance of power assist without being too sensitive. "We left the gear and the hoses exposed under the dash," says Steve. "So when you look at the car you see the mechanical parts of it."

Both the pitman arm and the steering arm used at the left front wheel were built in the Moal shop by Jack Friedland, the man responsible for all the precision machining on the car. Like many of the other parts on the car, the steering arms and linkage went through various stages and mock-ups before the arms and link were actually fabricated. The front arm is one Steve is especially proud of, explaining, "Jack made that arm from a solid chunk of stainless steel based on the design of an arm that a friend of mine, Jackie Howerton, makes. I think it's alright to use someone else's design as long as you acknowledge where the design or the inspiration came from."

Nearly all the parts on this roadster are hand made, even the upper mounts for the front shocks. Tim Remus

Other small parts, some of them chassis parts, were fabricated at about this same mid-point in the car's construction. Though the position of the front axle was already determined and the radius rods were already located along the side of the frame, the upper shock absorber mounting had never been fabricated and installed.

Once Steve knew where the upper shock mount would go, he started the process with a complete cardboard mock-up. When the mock-up fit, the dimensions were transferred to sheet steel

Once in place the shock mount looks very much like an integral part of this car. Michael Dobrin

Here you can see how the nose is still being developed. Louvers in lower panels are designed to allow hot air to exit the crowded engine compartment.

Steve strongly recommends pushing the car outside periodically for a good looking over and a photo session. It's a good way to check the development of the car. Michael Dobrin

This tubular framework for the nose is formed from sheet chrome moly. The idea was to make it strong yet very light. Michael Dobrin

and then the parts could be cut out and welded into a small structure. Like many of the parts on the car, what could have been just a simple, utilitarian triangle of metal turned out, instead, to be an arched and ventilated shock support that looks very much like part of an old sprint car.

Another example of the detail work that went into this car was provided one day when I stopped by to take some photos. Steve and Jack were working on the battery hold down. A"T" shape in cross section, the boys first cut a piece of flat stock long enough to reach across the battery. Next, another piece of flat strap nearly the same length and ventilated by half-holes, was welded to the first in order to give the piece some strength. Each piece had to be conceived, cut from strap, ventilated on the mill, and then welded together. Next the small piece of craftsmanship would go out for nickel plating. Nickel plated (not chrome) applied right over the raw steel with only minor polishing so some of the texture of the metal is sure to come through.

As the car started to come together other mechanical parts took their rightful place. In front, a pair of ventilated rotors and aluminum hubs were slid onto each spindle and a four piston Wilwood caliper was bolted to the bracket. The front and rear brakes are operated by a Wilwood twin-master brake assembly that allows the builder or driver to adjust the front-rear brake bias. The rear brakes too are discs squeezed by another pair of Wilwood aluminum calipers.

Next to the twin brake master cylinders is a third cylinder, this one used to operate the hydraulic clutch. "We used a hydraulic throwout bearing assembly from McLeod that puts everything inside the bell housing," explains Steve. "We used this style because we have no room for an outside slave cylinder, that's right where the driver's feet are."

After all his years of repairing exotic cars and building hot rods from scratch, Steve has learned a few very important tricks that may not be general knowledge. Like taking time to roll the partially competed project outside the garage periodically to check the progress and the proportions. Steve calls it the ability to see the trees for the forest: "You get so close to the project when you're in the shop that you can't really see the car. It's a good idea to roll them out periodically. I've discovered problems when I've done this in the past, and made significant changes to the car as a result. It's a case of stopping occasionally to step back from the car and make sure you're going down the right path."

It was after one of these little "review" sessions that Steve decided he really did like the third version of the car's nose. That meant he could construct an inner framework for the nose, an important part of the car's structure. Not only does it anchor the nose, it's also where the front of the hood comes down.

Though a heavy structure made of thick-wall moly tubing might have been the logical choice, Steve had different ideas. Using a sheet of .050inch 4130 chrome moly he started by making a box, shaped to fit the contours of the inner nose. To make it light and help it fit in harmony with the other pieces on the car, the sheet-metal box was lightened with a series of holes. Steve designed the tubular structure so it would bolt to the frame, instead of just welding it in place, for at least one very good reason. "That top bar needed to be removable to make it easy to install the engine. The box and the nose come off very easily which improves your angle of attack as you install the engine. We're trying to make the car pleasing to the eye yet make it as serviceable as possible."

The framework for the nose also supports the hood and provides structure for the whole front of the car. Michael Dobrin

After the nose was "approved" the rest of the body skin could be shaped and set in place. The side panels that run along the frame between the nose and the cowl are stepped so the hood sides overlap them at the upper edge. And to ensure that the air passing through the radiator has someplace to exit (remember, this car will have a full belly pan) well-known fabricator Bob Monroe punched a long series of louvers all along the length of each panel.

"At this stage we had the chassis pretty well finished," explains Steve. "Now we had to figure out, how do we cover the body, how do we get around all those linkages."

The framework for the door skins is developed until the shape is complete, then it's cut loose from the surrounding metal to become a functioning door. Michael Dobrin

A little farther along, the door framework is hinged and nearly ready for the outer aluminum skin. Steering wheel is a simple mock-up wheel. Michael Dobrin

Steve applies flux to a panel prior to welding. The idea is to create individual panels small enough that they can be handled by one man, and join those into an unbroken skin. All the panels are formed on the English wheel. Michael Dobrin

Most of the panels are formed from .050inch 3003 aluminum. This is tail section number one, which was later scrapped. Michael Dobrin

Once the basic shape of the cockpit was completed to his satisfaction, Steve went ahead and finished the framework for each door. It's interesting to note that most of the framework for each door was finished before being cut loose from the rest of the tubing surrounding the cockpit. Hinges and flanges were then added to each door. The outer skins were formed and attached to the framework last.

At about this same time work on the tail section commenced in earnest. Following the renderings done earlier Steve and Jimmy carefully formed the side panels and the trunk lid. Each panel was carefully set in place and gas-welded to the one next door. The work of forming the body panels was almost done and there was only one problem.

The problem was with the car's builder and designer. As he explains it now, "the tail of the car just didn't do it for me. I tried modifying the car with different taillights but finally I had to admit that it was just too damned short. I had to have a lot of board meetings with myself over the tail end of the car but eventually I cut it off and started over."

"I think the point is, it's OK to make a mistake. I'm willing to say I did it over. I'm proud of that. The car factories have warehouses full of parts they never used. You can't redo every little part, but if something is incorrect you have to own up to that fact.

The new tail is a full six inches longer and it looks much, much better. The added length allowed Jimmy to develop those wheel arches. I think now the tail of the car is one of its distinctive features. It gives the car 'The Look.' The new tail made us real happy with the car. I can sell it to the customer now because I wish it were mine."

Before Steve could deliver or "sell" the car to Tim Allen, there were till a few more details to sort out, a few more parts to fabricate from scratch. Like the wheels, which are being cut from solid aluminum as this book goes to press.

"In the shop for the mock-ups we put on a set of Real Wheels designed by Eric Vaughn", says Steve. "I really like those wheels, but we decided to offer this car a wheel that would be part of its own personality. The wheels are so important to a car, we felt these should be unique.

The new wheels measure seventeen inches in back and sixteen in front. Lil' John Buttera, the man who showed Boyd how to make billet wheels, fabricated the centers from billets of 6061 T6 aluminum with design input from Chip Foose and Steve.

In place of the mock-up engine seen in these photos the car will receive a new 351 Ford engine with GT-40 aluminum heads, all assembled and blue-printed by Cub Barnett. The steering wheel seen in the photos is also a mock-up. The final wheel will feature four spokes cut from billet aluminum supporting a laminated wooden rim built by Steve's father, George Moal. Also to come is the paint job, done in the paint booth at the Moal Body Shop in black urethane.

When it's all done, you might be tempted to ask, what is it? The simple answer would be 'cool.' A more thoughtful answer would describe the car as part coach-built Ferrari, part Indy roadster and part American hot rod. Ultimately the Tim Allen roadster is testimony to the

All the aluminum panels are joined by gas welding. Michael Dobrin

Wheels seen here are not the final choice. One-off 17 and 16 inch billet wheels will eventually be used. Bench in foreground is used to stand on to better asses the shape of the car. Like running it outside, the bench provides the proper perspective. Michael Dobrin

The finished steering arm is the work of Jack Friedland based on a design by Jackie Howerton. Michael Dobrin

The second, extended, rearend is fully six inches longer than the original. All of which helped provide Jimmy Kilroy with enough room to develop what Steve calls the "sweeping wheel arches." Michael Dobrin

A mostly completed roadster - with the new tail section - waits in the shop for final sheet metal work. The finished car will roll on four, one-off wheels, measuring seventeen inches in back and sixteen in front. Michael Dobrin

coach-building skills and design abilities of one man named Steve Moal and his very talented team of craftsmen.

Asked about the people who worked on the car, Steve is quick to share the glory and discuss his own responsibility. "I think when a person takes on a job like this a series of responsibilities come along with the work. We have a responsibility to the client of course. To produce something he likes, in a timely fashion at a reasonable cost. And then there's the responsibility of high quality engineering and the style of the car. But the largest responsibility is to acknowledge the input and skills of all the talented people who worked on the car. I must make sure people are aware that these cars are done by more than one person. The sign on the building has my name on it, but I certainly don't do all the work. I like to think that one of my talents is an ability to recognize the talents of others and take enjoyment from their high quality work."

If you ask Steve for his parting comments on the car, he explains again that he and his crew are "very happy with the car." And next he throws out a challenge to other dreamers, builders and fabricators. "My hope is that our work inspires people. Inspires them to do something better than what we've done."

Steve Moal and Jimmy Kilroy with the nearly finished Tim Allen roadster. Michael Dobrin

Front view of the nearly finished car shows the third and final version of the nose. "It's OK to say I did it over, and did it better." Steve Anderson

Chapter Eleven

Built on a Budget

Eric Builds an Early-Style Model A

Proud to be an amateur

If the roadster seen in Chapter Ten is a car professionally crafted by hand, the owner of the Model A describes his hot rod as, "a car that can be built by an amateur." Eric Aurand works as a graphic artist at Chassis Engineering. And while Eric's Model A project benefits from his association with Chassis Engineering, he did bolt the car together himself, much of it with help from his Father in the garage alongside his house in North Liberty, Iowa.

A Model A makes a good project for a a beginning hot rodder or anyone on a budget. A large number of original and reproduction parts are available at reasonable prices. Though it requires less money, a project like this still requires patience, an eye for detail, and a willingness to see the project to the end.

Anyone who feels that hot rods have gotten too expensive for the average guy or gal should pay close attention to this very simple, very functional hot rod. This car is a rolling testament to the fact that a neat street rod can be had for less than ten thousand dollars. Eric explains that a perfect street rod is not what he's trying to create here. "I want the finished car to feel like an old restoration, a car that someone cares for and maintains on a day by day basis. I'm using Rustoleum paint on the frame and doing all the body work myself for that reason."

A long-time automotive enthusiast, Eric goes on to express his feelings for the kind of car he's building here. "The magazine editorials blast the mail-order hot rods. I disagree. If you have to fabricate everything the project gets very expensive and you may end up with rolling stink. Anyone with a few skills can duplicate this car for seven or eight thousand dollars."

Eric's project started with a Model A Sport Coupe body purchased in South Dakota for the grand total of five hundred dollars. "That's a pretty typical price," explains Eric. "For the same money I could have had a five window coupe but I prefer the sport coupe because it's a somewhat unusual body style." Eric's decision to retain stock body dimensions, including an un-chopped top, is a big part of what makes this a very affordable project that any enthusiast with only average skills and tools can build at home.

There's a surprising amount of original tin still lying around in garages and barns across the US of A. "Original vintage tin is out there," says Eric. "If one is willing to pursue a few wild goose chases, listen to some old folks spin tales and poke through some dark and dirty barns." Eric adds, however, that there are a few things to consider before you drag that original tin all the way home. "Before purchasing an original body, be sure to check for evidence of collision damage, rust in vital areas, and the availability of any parts that are missing or damaged. You also need to check on the availability of a good reproduction chassis before buying the car and deciding to retain or discard the original frame."

The original fenders, running boards and splash aprons for the Model A were in poor condition, with plenty of cracks, rust and old repairs. Eric chose to leave all that iron oxide in South Dakota and pur-

Original tin still exists, you just have to dig a bit to find it. Eric found his Model A body hanging from the rafters of a barn in South Dakota.

After a thorough cleaning the bare body was set on an original frame to ensure all the body holes would line up and that the body was still square.

The new frame is fully boxed and features a front cross-member that is welded to both the inner and outer edges of the frame rails. It's important to do the mock-up with the water pump and fan in place.

The frame is set up on jack stands that reflect the rake that the finished car will have, before the degree wheel is set on the carburetor flange.

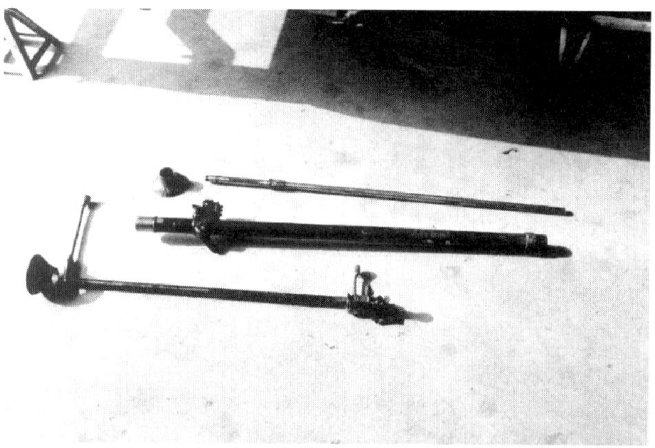

The steering column is from a 1940 Ford, minus the stock '40 steering gear. Without the steering gear the column needs a lower bearing, cut in this case from a piece of durlon (plastic bearing material).

The drop-out battery box is constructed while the frame is inverted. This location leaves the battery accessed from the bottom, only four bolts need be removed for its removal.

chased new pieces from Brookville Roadster company. "The steel fenders are pretty reasonably priced," says Eric. The whole kit only cost about fifteen hundred dollars. The firewall is from Bitchin, it's another example of the reasonably priced Model A parts that are out there and that helped to hold down the price of the project."

As mentioned in the Frame chapter, Eric chose to leave the chassis for his model A lying in the weeds next to the fenders. List price on the Chassis Engineering frame that Eric put under the Model A is only five hundred dollars. As Eric says, "for that kind of money it doesn't pay to drag the chassis back, strip it, repair the cracks and then box the frame rails." To digress for a moment, if the car were a 1934 Ford coupe then the cost of a reproduction frame goes up and the strength of an original frame does too so the decision on whether or not to retain the original frame must be made on a car-by-car basis.

The engine in the Model A is what you might call a full-house 303 cubic inch Oldsmobile V-8. Eric purchased the engine from friend who explained that the engine was originally used in a '40 Ford that was regularly campaigned on the drag strip in the '50s and '60s. In about 1965, after doing the annual rebuild of the Oldsmobile engine, the owners decided to install a small-block Chevy in its place.

The engine was never installed in a car again and thirty-some years later Eric purchased the well oiled V-8 complete with the two four barrel carbs, a high-lift Howard cam and Weiand finned aluminum valve covers. To quote Eric again, "I could have put a small-block Chevy engine in there pretty easy. You can find good used engines for two or three hundred dollars, a Ford engine might cost a bit more. But I wanted to use the Oldsmobile engine instead."

The transmission is a Saginaw four-speed, connected to the engine through a Speed Gems bell housing. Stock Oldsmobiles of the '50s used a single front engine mount and two, huge mounts on the stock bell housing. Eric converted the engine to a more conventional arrangement with two early aftermarket mounts that bolt to the large threaded boss located on either side of the Oldsmobile engine. These engine mounts sit on new rubber cushions and universal frame mounts, both from the Chassis Engineering catalog.

Back to the body

Once Eric had the body back home in Iowa, he stripped off all the original wood and canvas that make up the top on a sport coupe, and then gave the body a thorough cleaning. Next he set the body on an original frame, just to be sure it was straight and that all the body mounting holes lined up with the holes in the frame. The original body did contain some damage, and Eric had to repair a damaged quarter panel and some hacksaw cuts in the body where a previous owner began to convert the body to a pickup truck.

Though unfinished at the time we go to press, the interior wood and the canvas top will be replaced with a kit from Standard Auto Parts in Illinois. Though there is no instruction sheet Eric figures that with help from his wife Monica he can reassemble the wooden framework and stretch the new canvas into place.

The Chassis Engineering frame

The frame for Eric's home-built hot rod is a standard reproduction frame from Chassis Engineering. This frame matches the dimensions of the original, though it's much stronger. Like the stock Ford frame, the rails taper at the front and the back and the front frame horns are curved downward. Unlike the Ford frame, this one is fully boxed and comes with a stout K-member in the center of the frame.

As is the case with most quality reproduction frames from various companies, a number of options are available when ordering this frame. Eric chose the

After a bit of cleaning and sanding the new frame is ready for primer, followed by Rustoleum as the topcoat.

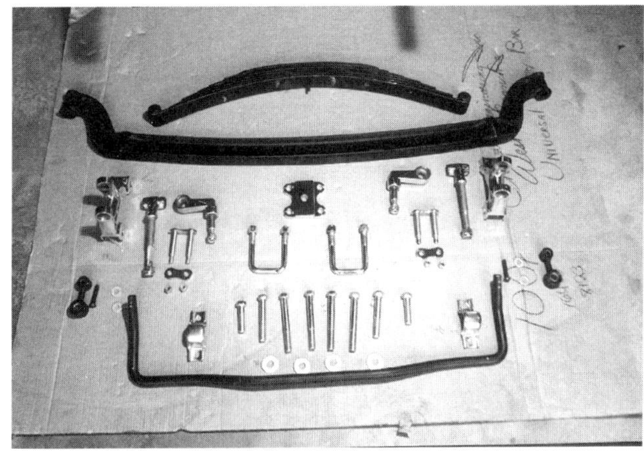

This is actually a frame just like Eric's under construction at the Chassis Eng. shop. It's a good idea to lay everything out and make sure it's all there.

Close up shows the back side of the spindle, before the grease zerks are installed. Lower bearing must be installed correctly and fit snug with shims that come with the kit.

The first job is to assemble the axle and spring into an assembly, then hang the four-bars in place.

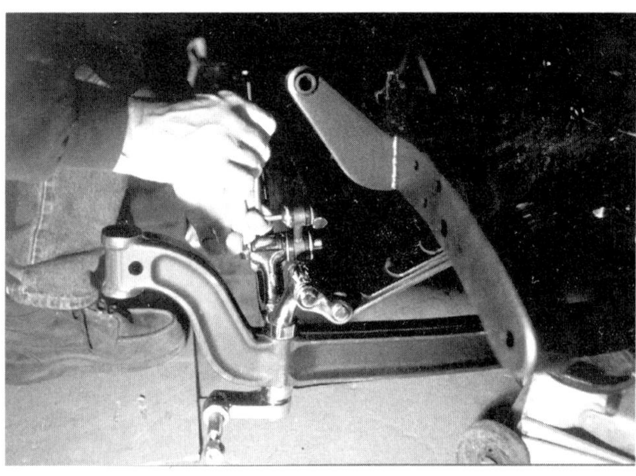

Chassis Engineering likes to install an anti-sway bar on a straight axle front end like this. The bracket connects the anti-sway bar to the forward four-bar bracket with the help of an extra long bolt and a special bracket.

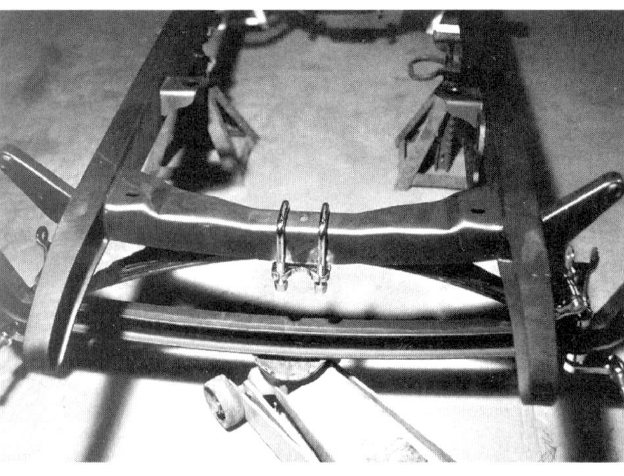

Simple chrome U-bolts mount the spring to the cross-member. Eric used a piece of welting between the top of the spring and the bottom of the cross-member. The cross-member in this frame is welded to both the inner and outer rail for strength.

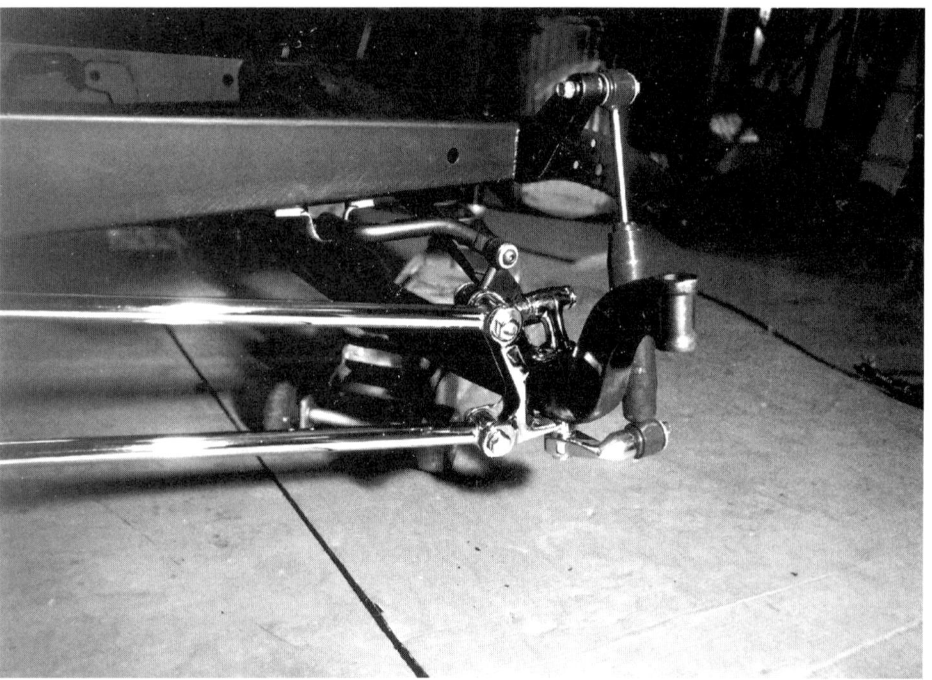

This shot shows the installed four-bars, anti-sway bar, bat wings and the shock. Lower shock bracket or "perch" is the one that uses a recess for the nut that bolts it to the axle, thus necessitating a thin-walled socket for installation.

triangulated four-bar suspension and a Ford nine-inch rearend, though he could have had a standard four-bar and an eight-inch Ford rearend, or mounting points and trailing arms installed to accommodate a modified Corvette independent rear suspension.

To start the initial mock-up of the frame, Eric set up the new chassis on four floor jacks, at the rake that the finished car will have. Then the engine, bell housing and rearend were lowered into place for a trial fit. As part of this mock-up process Eric temporarily installed a radiator and grille shell on the frame to be sure there was enough clearance for the fan. As Eric explains, "I also knew where the firewall would go, but the most critical clearance is at the front. Really, you have to put the engine where the radiator dictates and then do whatever you have to do to make the firewall fit. It also helped that we did all this in the Chassis Engineering shop, and the guys there know where most V-8s should sit in a Model A frame."

Once the engine and transmission were in the right location the two front engine mounts and the rear transmission mount were tack-welded into place. As shown in the photographs, the engine is installed into the frame at an angle of two to three degrees down at the back. Eric measured the angle with the protractor on the carb-mounting surface and then checked it again at the tailshaft because some carb-mounting flanges are tilted one way or another to compensate for the angle of the engine in an O.E.M. chassis.

A number of factors need to be considered before the final engine position is determined. As Eric explains, "The location of the steering box as well as the column should be decided upon before the engine and transmission mounts are finalized. I chose a Vega steering box and a column from a '40 Ford." Eric chose the Ford column because, as he explains, "it fit the theme of the car." The '40 Ford steering wheel will be used as well, though again Eric warns builders to check the wheel diameter and column location in the car, "so you aren't banging your knuckles on the door or dash every time you turn a corner." The other thing to be considered at this mock-up stage is the routing of the exhaust.

Once the position of the engine and transmission mounts are known Eric could pull the engine and transmission back out of the frame so the suspension mounts can be final-welded onto the rails. The Chassis Engineering crew already has a template to correctly locate the brackets needed for the triangulated four-bar linkage, and these are welded in place while the frame is in the jig. The location of the front mounts for the four-bar linkage to be used up front is known as well, and these are welded on at the same time.

The Model A is a small car, and the decision was made early to hang the battery underneath the frame. The battery box, with a drop-out provision, is also constructed at this time while the frame is inverted in the shop. A narrowed Ford nine-inch rearend is Eric's rearend of choice. Before sending it out to Lake City Rod and Custom in Watertown S.D. Eric carefully cut the spring pads off and then had the housing sand blasted. The

The rear suspension in Eric's Model A is a triangulated four-bar kit from Chassis Engineering.

You can see both the upper, angled, links that prevent lateral movement; and the parallel lower links. Brackets for four-bar links were added after the nine-inch was narrowed.

The finished frame with fresh paint is ready to be rolled into the garage.

Before the body can be set down on the frame welting must be applied. This is original Ford-type welting. A piece of sharpened pipe is used to cut holes where body bolts go through.

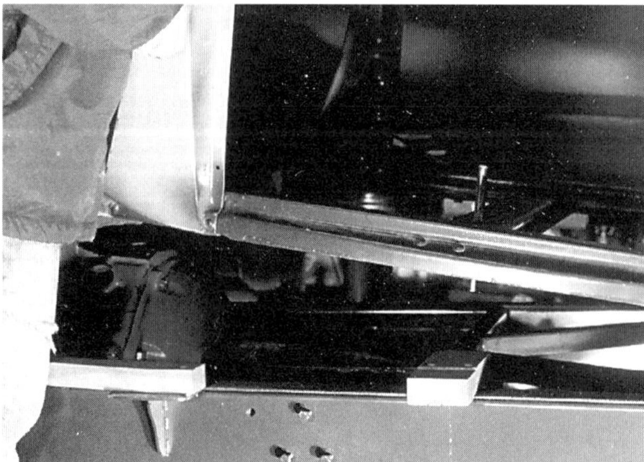

Once the welting and wood mounting blocks are in place the body can be set down for a test fit. Splash aprons and fenders will be fit after all body mounts have been properly fit and adjusted.

correct triangulated four-bar brackets were later installed in the Chassis Engineering shop. As always, any cutting and welding done to the housing is done carefully so as to avoid warping the metal.

Install the suspension

Once the suspension brackets and engine mounts are welded in place Eric cleans up the frame and prepares it for paint. Rustoleum, available at your local hardware store, is the durable and low-tech paint Eric used for the chassis and much of the associated hardware. Once the paint is dry Eric can begin the installation of the front and rear axles and suspension components.

"Before putting on the front axle it's a good idea to lay out all the parts on the floor," advises Eric. "just to be sure you've got all the necessary parts, brackets and bolts. Then it's pretty simple to just bolt together the axle and spring into an assembly. Next we lifted one end up, attached the four bars, and did the same thing on the other side. Assembly of the front axle and four bars is a very simple and straightforward operation. I did have to buy a thin-wall socket from Snap-On to tighten the nut for the lower shock perch."

The kit from Chassis Engineering includes an anti-sway bar. Some builders leave the anti-sway bar, drag link and tie rod off until later, to leave more room during the engine installation.

With the axle in place it's time to install the spindles and kingpins. Though a pretty simple operation, Eric points out again that there are pitfalls for the inexperienced builder. "Our kingpin and spindle kits come with the bushings installed in the spindles *and reamed to the correct size for the kingpins.* If the bushings aren't reamed to size you have to take them to your local automotive machine shop and have the operation performed. Also, the lower ball bearing, the one that actually supports the car's weight, has an open and a closed side. The open side should face down so when the kingpins are greased the grease goes up into the bearing."

With the spindles in place Eric installed the brake caliper brackets next. Because the brackets absorb so much torque it's important to use the high-quality bolts that come with the kit, and to be sure the right bolt goes in the right hole (some are longer than others). The brake kit used on the Model A uses mid-size GM calipers from cars like the Cutlass and Regal. The rotors measure 10-3/4 inches in diameter and come from the Mopar side of the aisle. Eric obtained his from a local junk yard, taken off a Plymouth Volare, though a variety of rear-wheel drive Chrysler products use the same rotor on the front end.

The factory provides a minimum thickness specification for any rotor. When buying used it's important to be sure the rotors have enough material remaining that they can be turned and still meet that minimum thickness requirement. Wheel bearings and races should be inspected too, re-packed with a good bearing packer and installed with a new wheel seal. The tightness of the wheel baring nut is rather critical and Eric followed the torque recommendations in a factory service manual for tightening the spindle nuts on his front spindles.

Installation of the rearend came next. Installing the complete and painted housing is pretty simple, as

Construction adhesive is used, in addition to screws, to secure the new wood into position. The channel had to be spread slightly, clamps pull it all back into place.

One of the nice things about building a Model A is the fact that original "wood kits" are still available. Eric removed the old wood, cleaned the channel, and painted the metal.

long as you've got a floor jack and a friend to help wrestle it into place. In Eric's case, the rear wheel cylinders, brake shoes and hardware were all replaced with new components before installation. The final part of this operation is the installation of the Aldan coil-overs with the 250 pound springs.

Mock-up the body

Once the chassis is completed, Eric could begin the job of repairing the body and installing it on the frame. Most older cars used a welting between the body and frame and the Model A is no different. Though welting with peel-off adhesive is available, the product used here is a simple roll of original-type material. Eric glues it to the top of the frame rails only in places where the welting must be cut to conform to the rail. For the body-mounting bolts a hole

Like the interior wood, nearly every Model A body and support bracket is available as well. These subfloor extensions are clamped together and aligned with a piece of the body block, before being welded with the wire-feed.

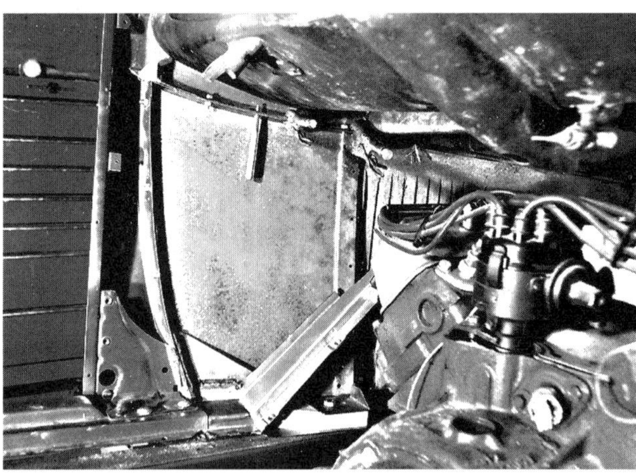

Because the gas tank is part of the cowl, it must stay in place until the cowl and all related supports are correctly fit and in place, then the bottom can be cut out. Subfloor extensions are used to support the bottom of the cowl where it meet the floor.

Once the body is sitting square on the frame the steering, firewall and fenders can be test fit.

is punched in the welting with a piece of tubing and a hammer.

With the welting in place it's time for the first trial fit of the old Model A body. Wooden body-mounting blocks are set in place on top of the welting before the body is set on. These blocks fit snugly up into the concave subfloor along the side of the body and end up flush with the bottom of the body once everything is set in place.

Now the sport-coupe body is set down on the frame. Eric mounted the body on another stock frame earlier, and was thus assured the body was square and would most likely fit this frame without many problems.

The cowl side panels are bolted in place on the body next, to see how they fit and to visualize exactly where the firewall will go. A couple of bolts are inserted through the body-mounting holes to prevent the body from sliding off the frame. Now it's time to mock-up the steering and firewall and test fit the fenders and splash aprons.

Among the items being test fit at this point is the dash, a trimmed unit from a '32 Ford. The test fit can't be done completely however, because the Model A gas tank (which functions as the dash as well) is still in place.

With the body sitting square on the frame, Eric and his father go ahead and replace the rusty subfloor extensions, located at the bottom of the cowl on the sides. These subfloor extensions have to be clamped into an assembly, using the body-mount wood block for alignment, before being welded together and attached to the cowl side panels.

Next the cowl sides are fitted and bolted back onto the rest of the body. Eric notes that you have to be patient before cutting out the bottom of the gas tank. "If the bottom of the gas tank is to be cut out to provide under-dash space for wiring and all the rest, this should be done last as the tank's structure helps to align the cowl.

Much of the wood used to reinforce the old body is replaced at this time. The first step is to treat the new wood that comes in the kit with a preservative and allow that to dry before beginning the installation. Once the original wood is out of the way, Eric can wire brush the channel and apply a coating of Rustoleum.

Once that paint has dried the new wood can be pushed into position. As Eric explains, "The door jamb may have to be spread slightly to get the new piece in past the old counter sink indentations in the door-side of the metal. I put a little construction adhesive in place on the quarter-panel side of the wood and used clamps to draw the metal jamb back into position. Then we replaced the screws and bolts."

Now it's time to install the running boards and splash aprons. Eric starts by loosely bolting on the support brackets for the running boards. With the running boards bolted to the supports, they can be tightened to the frame. In Eric's case one of the support brackets had to be heated and bent to fit properly. It turned out later that the kit contained two right-side brackets so the angle was wrong when one of

New running board supports are bolted to the frame but left loose until later.

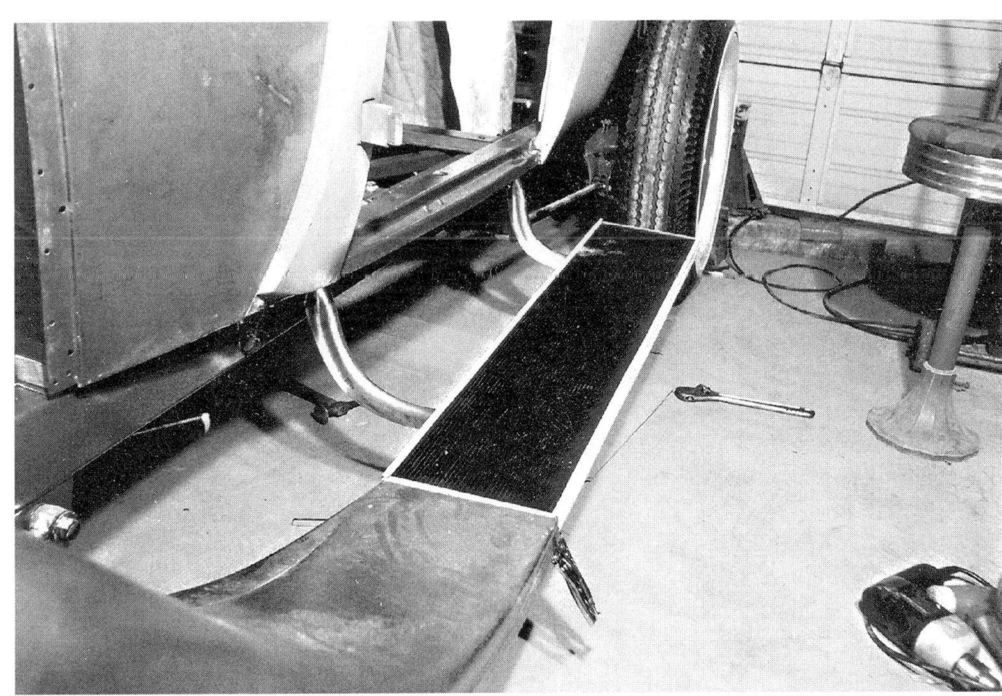

The running board is fastened to the supports then the supports are tightened to the frame. Fender is clamped to the running board for alignment.

Hard to believe, but not everything lined up just perfect. Here a small torch is used to improve the fitment of the support brackets.

This close up shows how the splash apron fits under the fender and on top of the frame welting. Much trial and error is required to get all this sheet metal to fit correctly.

those was used on the left. The rear fenders are bolted to the body and the running boards next.

"The front fenders are placed on the frame now," explains Eric. "With the headlight bar in place I had to align the fenders with the center-punch marks on the top of the frame rails (not all aftermarket rails have these marks). Then we took the fenders off again so we could drill and tap the holes for the fenders. The Chassis Engineering frames are marked for both the '28 - '29 fender holes, and the '30 - '31 hole locations. It's important to use the correct hole location. "Some variation on fender stampings and pre-punched fender holes may require slotting of the hole in the fender for proper alignment. We had the fenders off and on a couple of times just to be sure of the location and the alignment."

Eric goes on to explain, "Now that the running board has been fit and the front and rear fenders clamped in place to check their location, the running board and rear fender must be removed to facilitate the installation of the side splash apron. This apron slides inside the front fender and sits on top of the frame welting but under the wooden body blocks. It may help to have a hood at this point, in order to check the radiator-shell to cowl distance, and to line up the hood shelf holes in the splash apron."

In order to install the splash apron the rear fender and running boards must be removed and the body lifted up again.

Next comes installation of the Bitchin Products floor boards. Eric reports that all the parts fit really nice and that the installation is another straightforward operation. Eric is installing the column and Bitchin firewall as the book goes to press. Still ahead is the removal of the bottom half of the gas tank and installation of the '32 Ford dash.

The floorboard kit is fit after the body is mounted to the frame. This kit is from Bitchin, another example of all the quality parts available for a Model A.

Eric explains that in spite of the missing pieces, "The assembly of the new and used parts now forms a car. The really big job that lies ahead is in the finishing work: Firing the engine for the first time (it has run recently on a test stand), and disassembling the body for body work and paint. Then comes installation of the electrical system and Painless wiring harness, the insulation and sound proofing, plus the top and upholstery. A large supply of nuts and bolts, sand paper and plenty of time will be consumed over the winter months in order to take it out for a spin in the spring. A heated garage and a long Midwestern winter provide the impetus for such a project. For anyone doing a first-time hot rod or antique car, I think it's important to pace yourself and remember to have fun. A great leaning experience and sense of accomplishment are many times more valuable than the hard product."

Eric ends with a few final word of thanks to Pinetree Welding, in Baraboo Wisconsin and Yogi's Hot Rod Parts in Calamus, Iowa.

After a thousand test fits the old Model A looks like a car. Now comes the job of blowing it all apart again for paint and upholstery.

Sources

Bitchin Products
9392 Bond Ave.
El Cajon, CA 92021
619 443 7703

Roy Brizio Street Rods
263 Wattis Way So
San Francisco, CA 94089
650 952 7637

Brookville Roadster
718 Albert Rd.
Brookville, OH 45309
937 833 4605

Chassis Engineering
119 No. 2nd.
West Branch, IA 52358
319 643-2645

Concours West Industries
300 Dwight Rd.
Castle Rock, WA 98611
360 274 3373

Currie Enterprises
1480B. N. Tustin Ave
Anaheim, CA 92807
714 528 6957

Downs Manufacturing
715 N Main St.
Lawton, MI 49065
616 624 4081

Dutchmen Motorsports
PO Box 20505
Portland, OR 97294
503 257 6604

Ford Motorsport
17000 Southfield Rd
Allen Park, MI 48101
810 468 1356

Friendly Chev.
7501 Hwy 65 NE
Fridley, MN 55432
651786 6100

Gibbon Fiberglass
132 Industrial Way
Darlington, SC 29532
843 395 6200

Lake City Rod & Custom
14 3rd St NW
Watertown, S.D. 57201
605 882 5700

Heidt's Hot Rod Shop
5420 Newport Dr. #49
Rolling Meadows, IL 60008
800 841 8188

Metal Fab
1453 91st Av NE
Blaine, MN 55434
651 786-8910

Metro-Matic Transmission
4030 Hoffman Rd
Wite Bear Lk, MN, 55110
651 426 9122

Mike Adams Palm Beach R-S
1492 53rd St.
West Palm Beach, FL 33407
561 844 4017

Mopar Performance
248 853 7290
800 348 4696

Phoenix Transmission Products
922 Fort Worth Hwy
Weatherford, TX 76086
817 599 7680

Pure Choice
2155 W Acoma Blvd.
Lake Havasu City, AZ 86403
520 505 8355

Standard Auto Parts
3813 CulDeSack
Peoria Illinois, 61604
309 637 6668

Wheeler Racing Engines
10500 Nasau St.
Blaine, MN 55434
PH: 651 786-3278